The Trouble with Technology

The Trouble with Technology

Explorations in the Process of Technological Change

Edited by
Stuart Macdonald
D. McL. Lamberton
Thomas Mandeville

St. Martin's Press, New York

© Don Lamberton, Stuart Macdonald, and Tom Mandeville

All rights reserved. For information, write:
St. Martin's Press, Inc., 175 Fifth Avenue, New York, NY 10010
Printed in Great Britain
First published in the United States of America in 1983

Library of Congress Cataloging in Publication Data
Main entry under title:

The Trouble with technology.

    Includes indexes.
    1. Technology—Addresses, essays, lectures. I. Lamberton, D. M. (Donald McLean), 1927- . II. Macdonald, Stuart. III. Mandeville, T. D.
T185.T76 1983   338.4'76   83-10961
ISBN 0-312-81985-4

Contents

Acknowledgements	vii
List of Contributors	ix
List of Figures	xi
List of Tables	xii
1. Introduction	1

Part I: Towards conceptualising the process of technological change

2. Technology in the evolutionary process *Kenneth E. Boulding*	4
3. The machine: icon of economic growth *Peter Mathias*	11
4. Technology beyond machines *Stuart Macdonald*	26

Part II: The economic theorist's dilemma

5. Can we explain technical change? *Arnold Heertje*	37
6. A conceptual framework for modelling the role of technological change *Duncan Ironmonger*	50
7. The accumulation of intangibles by high-technology firms *M. Teubal*	56
8. Information economics and technological change *D. McL. Lamberton*	75

Part III: Diffusion, technology transfer and trade

9. Theoretical approaches to the analysis of the diffusion of new technology *P. Stoneman*	93

Contents

10. On the adoption of technological innovations in industry: superficial models and complex decision processes — 104
    Bela Gold

11. Technical advance and trade advantage — 122
    D. K. Stout

12. Trade, technology transfer, and development — 134
    Meheroo Jussawalla

13. The technology transfer process in foreign licensing arrangements — 155
    Lawrence S. Welch

Part IV: From employment to policy

14. Information technology and employment levels — 169
    Thomas Mandeville and Stuart Macdonald

15. Trade unions and technological change — 178
    John Corina

16. The trouble with techno speak — 193
    Ian Reinecke

17. The difficulties of national innovation policies — 202
    Roy Rothwell

Author Index — 217

Subject Index — 221

Acknowledgements

The Editors are grateful to the Department of Economics of the University of Queensland for providing facilities necessary for the production of this book. The secretarial staff of that Department has been particularly helpful, often under trying circumstances. Ms Geok Latham was responsible for co-ordinating and organising much of the rather tedious detail that a work like this entails—a frustrating task which really deserves much more than mere acknowledgement.

Contributors

KENNETH E. BOULDING is Research Associate and Project Director in the Research Program on Political and Economic Change at the Institute of Behavioral Science and Distinguished Professor Emeritus of Economics at the University of Colorado.

JOHN CORINA is Foundation Professor of Industrial Relations at the University of Sydney.

BELA GOLD is William Umstattd Professor of Industrial Economics and Director of the Research Program in Industrial Economics at Case Western Reserve University, Cleveland, Ohio.

ARNOLD HEERTJE is Professor of Economics at the University of Amsterdam and currently Visiting Professor at the Department of Economics, University of California at Berkeley.

DUNCAN IRONMONGER is Reader in Economics and Deputy Director of the Institute of Applied Economic and Social Research, University of Melbourne.

MEHEROO JUSSAWALLA is Research Associate in Economics at the East-West Communication Institute, East-West Center, Honolulu.

D. McL. LAMBERTON is Professor of Economics in the Information Research Unit, Department of Economics, University of Queensland.

STUART MACDONALD is a Senior Lecturer in the Information Research Unit, Department of Economics, University of Queensland.

THOMAS MANDEVILLE is Research Officer in the Information Research Unit, Department of Economics, University of Queensland.

PETER MATHIAS is Chichele Professor of Economic History at the University of Oxford and Fellow of All Souls College.

IAN REINECKE is Publisher, Technical Group with Thomson Publications Australia.

ROY ROTHWELL is a Senior Research Fellow with the Science Policy Research Unit at the University of Sussex.

P. STONEMAN is a Senior Lecturer in the Department of Economics, University of Warwick.

D. K. STOUT is Chief Economist of Unilever and Visiting Professor of Economics at the University of Leicester.

M. TEUBAL is a Senior Lecturer in the Department of Economics at the Hebrew University of Jerusalem.

LAWRENCE S. WELCH is Visiting Professor at the Stiftelsen Bedriftsøkonomisk Institutt (Norwegian School of Management) in Oslo.

List of Figures

7.1.	Project profitability and interrelationships (electronic instruments firm)	64
8.1.	Information occupations as a percentage of economically active	81
10.1.	Alternative conceptions of innovation diffusion patterns	107
12.1.	OECD imports of manufactured products from developing countries by broad product groups, 1970-79	151
13.1.	An interaction perspective on technology transfer	164
15.1.	APEX strategy for the introduction of word processing and other office automation	185
17.1.	Industrial innovation—possible benefits	206

List of Tables

4.1.	Persons required per million dollars of annual output by selected Australian industries in 1990/91	28
7.1.	R & D profitability of the case firm's projects	63
7.2.	Direct and indirect profitability (electronic instruments firm)	68
8.1.	Typology of information occupations	82
17.1.	Classification of government policy tools	203
17.2.	List of reports used for analysis of national innovation policy formulations	205
17.3.	Industrial innovation policies—tactical objectives	207
17.4.	Industrial innovation policies—main policy measures	208
17.5.	Analysis of policy recommendations by type of tool	209
17.6.	Some contradictory policies affecting industry in the United States	212

1. Introduction

The trouble with technology started long ago. As Joseph Schumpeter pointed out:

> Long before the industrial revolution, people realized the obvious fact that machinery often displaces labor . . . [G]overnments and writers worried about this and labour groups and citizens' guilds fought against machinery, the more so because immediate effects of this kind are concentrated in time and place, whereas the long-run effects on general wealth are much less visible in the short run and much less easy to trace to the machine.[1]

The tragedy is that decision makers, be they members of the public, private or public corporations, or political actors, do not have ready access to even this basic wisdom, let alone a more sophisticated understanding of the very complex process of technological change. Lacking such understanding, almost inevitably it seems, opinion is drawn towards the polar camps of optimists and pessimists. Although the truth lies in between, Schumpeter was undoubtedly correct when he taxed economists with preoccupation with 'fighting the public's propensity to attend too exclusively to temporary phenomena', because the economists 'attended too little to temporary phenomena themselves'.[2] Such preoccupation has been supported strongly by long traditions of static equilibrium theorising by economists and self-interest on the part of the other economic actors.

Two general points on the side of pessimists deserve special consideration. First, as Manfred Stanley argued: 'The pessimistic case seems . . . on the face of it more informed by humanistic sensitivities'.[3] While the economist *qua* economist may not wish to pursue this point, he has difficulty evading the second: the traditional arguments in favour of superiority of the market economy do not extend to cost minimisation, responsiveness to change, and innovativeness.[4] These limitations have an important bearing upon the creation of technology, its adoption and diffusion, and the perception of and response to its effects. Arrow does well to remind us that the Manhattan Project (the development of the atomic bomb) would have been regarded as:

> a supreme example of the superiority of a private enterprise system had it in fact been done by private enterprise. The project represented a degree of sophistication in resource allocation in a very, very challenging set of circumstances and vast uncertainties . . . [It] was full of organizational innovations

—especially the idea of parallel developments of alternative possible solutions—and many of the lessons of that experience seem to have been completely ignored in almost all subsequent research and development work, whether by the government or by private industry.[5]

Before turning to comment on the scope of papers in this book and their shared emphases, it is useful to reflect that the trouble with technology extends far beyond the economist's professional competence. In a short, provocative essay, 'Thoughts on art and technology', Jane Livingston considered the possibility that there could be a valuable interchange between artists and members of the corporate industrial society. Attitudes were pivotal.

One of the fundamental dualisms inherent in the question of technology's uses in a humanist context has to do with the conflict between the belief that, in a word, technology *is* the metaphysics of this century, and therefore has to be accommodated from within, and the view that technology is somehow self-perpetuating, implacable and *essentially* inhuman, and that therefore humanists and artistic endeavour must function separated from it and even in opposition to it. Nearly all the positions taken by artists and by their scientific counterparts with respect to the art/technology relationship are conditioned by one or the other of these antithetical beliefs.[6]

Is there reason to hope that economists can move beyond the confines of their traditional attitudes and models? Can they achieve a view between such antithetical beliefs? The papers in this volume constitute a powerful criticism of orthodoxy with its neglect of the process of technological change. The contributions reflect the growing tendency to seek to treat technological change as endogenous, to seek to understand the ways in which technology is created by and conditioned by the functioning of the economic system and is yet capable of changing that system. The indebtness to Marx and Schumpeter is obvious. A part of this new effort to cope with the trouble with technology is apparent in renewed emphasis upon organisational or institutional aspects. Concepts such as organisational capital and intangibles are introduced to analyses that hitherto attempted to rely upon the bare traditional concepts of capital and labour. In each case a form of investment is involved: investment that allows analytically for the influence of past events or present decisions.

Technology is one kind of information, perhaps the most important kind and even the most important resource of society. The varied but interconnected elements of what is coming to be called information economics constitute in part an attempt to elucidate the full implications of escaping from 'perfect knowledge' models and treating information as a resource. Its more important contribution is that having left the artificial models of 'perfect knowledge' or even 'given uncertainties', it is no longer possible to avoid treating the organisation as a variable. Inventing the future means inventing both new machines and new ways of organising social and economic activity. And yet to rid economics of people, both consumers and entrepreneurs, and institutions has been seen as the crowning achievement of economic theory!

Small wonder that these contrasts have brought even such a distinguished economist as Joan Robinson to feel that economic theory has 'come to pieces'.[7] She is reported as holding the view that 'any long-run theory seems ... to be quite impossible, because it depends upon the development of technology'.[8] While it is true that technology undermines static ideas of capital and makes measurement seem virtually impossible (can current investment in information be measured? can the stock of information capital be measured?), economists should not despair, but should be prepared to concede that many widely accepted conclusions are very sensitive to the assumptions made about information. In humble fashion they should get on with research into the economics of internal organisation and its informational aspects because in that way they may eventually reunite macroeconomics and microeconomics and achieve better policy recommendations—recommendations that no longer focus on the 'optimal' quantities of capital and labour alone, but ask if organisations are obsolete and need to be redesigned. The explorations in this book are offered in that spirit and with that hope.

Notes and references

1. Schumpeter, J. (1954), *History of Economic Analysis*, New York, Oxford University Press, p. 680.
2. Ibid., p. 681.
3. Stanley, M. (1978), *The Technological Conscience*, Chicago, University of Chicago Press, p. xi.
4. See, for example, Nelson, R., and S. Winter (1982), *An Evolutionary Theory of Economic Change*, Cambridge, Mass., The Belknap Press of Harvard University Press, esp. Ch. 15.
5. Arrow, K. (1979), 'The limitations of the profit motive', *Challenge*, 22 (4) : 27.
6. Livingston, J. (1971), 'Thoughts on art and technology' in Maurice Tuchman (ed.), *A Report on the Art and Technology Program of the Los Angeles County Museum of Art 1967-1971*, Los Angeles County Museum of Art, p. 43.
7. Davie, M. (1983), 'The economist fires a salvo ... at economics', *National Times*, 6-12 February, p. 10.
8. Idem.

PART I

TOWARDS CONCEPTUALISING THE PROCESS OF TECHNOLOGICAL CHANGE

2. Technology in the evolutionary process

Kenneth E. Boulding

Evolution can be defined as ecological interaction and succession under conditions of constantly changing parameters. This process really started with the 'Big Bang'. It exhibits three major phases in our part of the universe: first, pre-biological evolution, which produced the elements, radiation of different wavelengths, the compounds, the galaxies, stars and planets in enormous variety; and second, biological evolution, starting with the introduction (nobody knows how) of DNA with its extraordinary capacity for self-replication and for organising the processes of production of organisms. Our own planet is the only example of this that we know about, and it would certainly be surprising if this were the only example in a large universe, even though the development of life and biological evolution is highly improbable and requires a very unusual physical environment. It is clear, for instance, that earth is the only object in the solar system which is capable of sustaining biological evolution. The third phase comes with the development of the human race with its extraordinary capacity for knowledge—that is, structures inside the organism which in some way map into the structures of the real world both inside and outside it. From this capacity comes the huge proliferation of human artefacts—ranging from the first eolith to the space shuttle—which originate like biological organisms in some kind of 'know-how', whether this is the genetic instruction in the fertilised egg that produces the organism of the human know-how coded in large numbers of human minds and other human artefacts, blueprints and instruction sheets that produces the car. Human artefacts are of three kinds. There are things, that is, material objects, of which we now have a much larger number of species than there are biological species. Then there are organisational artefacts, ranging from the family and the hunting band of the Palaeolithic to the Catholic Church, General Motors and the United Nations. Each human person is also partly a biological artefact from his or her fertilised egg and genetic structure, and in part a human artefact in terms of language, a good deal of behaviour, knowledge, valuations, know-how, and so on.

There is a rather curious illusion among those who might, perhaps a little unkindly, be called romantics. They believe that biological artefacts, and, indeed, physical and chemical artefacts (such as the rocks, the ocean, and the atmosphere, which have been produced by the processes of pre-biological evolution) are in some sense part of a majestic whole called 'nature', whereas human beings and their artefacts are not part of nature, are even hostile to nature, and are in some sense artificial. There seems to me no justification for this belief. The development of human beings and their artefacts is just as much part of the evolutionary process as the development of biological organisms. Indeed, human beings are biological organisms. Even before human beings, some biological organisms produced artefacts—beaver dams and termite nests, for example—though in comparison with human artefacts these are extremely limited. Once biological evolution had produced human beings, however, they were clearly part of the world ecosystem and their artefacts were also part of the world ecosystem, just as beaver dams are. I have joked that the car is a species just like the horse—but with a more complicated sex life. Biological artefacts—that is, organisms—contain the genetic know-how for producing a similar organism within the organism itself, or at least within two organisms of opposite sex in the case of sexual reproduction. The reproduction of human artefacts is 'multi-parental'. The know-how that reproduces them is not contained in the artefact itself, but in a very large number of other human beings, organisations, and material artefacts. The selective processes, however, by which the populations of all interacting species change is fairly similar in biological evolution and societal, which might be called 'post-biological', evolution. Whereas pre-human biological organisms for the most part interact unconsciously, humans interact consciously, and their artefacts depend very heavily on human valuations of them. Evaluation goes back a very long way in biological evolution, although much of it is genetically developed. Sexual attraction, parental care, selection of food, even the creation of pre-human artefacts like spiders' webs or nests, all involve evaluation. One could even argue, indeed, that chemical valency is a primitive form of valuation. Carbon 'likes' having four hydrogens, and CH_3 is a radical, being unsatisfied. In the human race, however, evaluation becomes overwhelmingly important as a determinant of behaviour, in terms both of going towards what we like and of going away from what we do not like.

The basic concept of an ecological system is that of a 'niche' which can be defined as an equilibrium population of a species. The definition of a species is not always easy, but we can leave that aside for now. The ecosystem is a system of interacting populations of different species, the interactions being very complicated, including co-operative, competitive, and predative relationships, food chains, territoriality, space and time competition, and so on. The rate of growth of any population is equal to the additions minus the subtractions, and these in turn are functions of the size of all the other populations with which it interacts. In a given environment, as the population grows, it becomes a little harder to add to it and easier to subtract. So it grows to the point at which additions equal subtractions and the population is stable, at least over short-term fluctuations. The equilibrium of a whole ecosystem is one in which all populations are at the level where they are stable. This equilibrium itself may be stable for relatively small disturbances. Take 20 per cent of a given kind of fish out of a pond and in a year or two the population will be back to where it was before, unless a disturbance creates some change in the parameters of the system. All ecosystems are likely to have empty niches—that is, species that would have an equilibrium population in the system if they existed. Australia clearly had an empty niche for rabbits in the nineteenth century, and for European-type humans. Empty niches can be filled either by mutation of some kind or by migration. If an empty niche is filled, this changes all the other niches; some, indeed, may shrink to zero and the species becomes extinct. Though ecological equilibria are always very temporary—if they exist at all in any exact form—the concept is useful.

It is important to recognise that there are two forms of genetic structure: one is 'biogenetic', which is DNA and all that, highly Mendelian in its operation. The experience of the organism rarely changes the structure of its genes. The other I have called 'noogenetics' (from nous), which is learned structures transmitted from one generation to the next by a learning process. Biogenetic processes create the potential for such learning in nervous systems and other structures. This potential can be realised only through a learning process in the organism itself. Some of this can be found at pre-human levels (bird song, for example), but again in the human race this becomes of overwhelming importance. Mutation in terms of human artefacts consists wholly of noogenetic evolution—that is, learning new forms of know-how. The crucial importance of know-how in both biological and societal evolution can be seen very clearly if we ask ourselves why there were no plastics a hundred years ago. The answer is that we did not know how to make them. Similarly, why were there no human beings ten million years ago? The answer is that the biogenetic structures did not know how to make them.

The niches of human artefacts are created almost entirely by human evaluations. This takes place through three major processes which I have called the exchange system, the threat system, and the integrative system. Cars exist primarily because there is a demand for them; people can afford them, and they can be produced at a price that people can afford. Nuclear weapons exist because there is a demand for them from national states, which can afford them because they tax people through the legitimated threat system of the law and cheat people through inflation. Australia, or any national state, exists because people believe it exists, and for no other reason. They believe it exists because of very complex historical integrative processes that make people identify with particular communities and not with others.

One of the real puzzles of human history is why change in human artefacts, at least to judge by their remains, was so extraordinarily slow in the 40,000 years or more of the Palaeolithic. Creatures who were genetically virtually identical with us continued apparently in the same state of culture without learning very much for literally hundreds of generations. The answer may lie partly in their short length of life. Life expectancy of thirty years gives very little time for acquiring new knowledge. All the learning activity has to be devoted to transmitting the existing knowledge from one generation to the next. It is with the development of agriculture that we begin to get something of a surplus. Life expectancy rises to perhaps forty years in the Neolithic, and for the first time in human history there is time to think, to plan, to change; there are opportunities to transmit new knowledge to succeeding generations. There was no empty niche for civilisations before agriculture; there would not have been enough food surplus to support it. Even after the development of agriculture it seems to have taken several thousand years for the niche of soldiers, priests, kings, bureaucrats, architects, builders, and so on to be filled, resulting from the excess food supply which the managers of threat, whether priests or kings, were able to collect from the food producer. Mutation takes time and the more improbable the mutation, the more time it is likely to take.

Whether we can separate something called 'technology' from the ongoing process of human learning and the production of human artefacts is an intriguing question. In a certain sense technology begins with the first human artefact, whatever it was. On the other hand, there is a certain break between pre-scientific artefacts, which come out of folk knowledge and often almost unconscious skill, and science-based artefacts, which come out of the extraordinary expansion of human knowledge which has been the result of the development of the scientific subculture in the last five hundred years or so. One can argue indeed that science-based technology was not very important before about 1860; it is tempting almost to place the Crystal Palace, the first steel-frame building, at the beginning of it. The so-called industrial revolution in England in the eighteenth century was built largely on development from mediaeval technology. Even the steam engine owed nothing to thermodynamics, nor did the railway. From about 1860, however, we do detect an almost exponential explosion of science-based technology, beginning perhaps with the chemical industry. The electrical industry in the 1880s caused an enormous transformation of the planet and created niches for innumerable appliances and artefacts. The skyscraper, a result of improvement in steel production and perhaps glass, again is in good part a result of chemistry and a little physics. Medical science, and still more perhaps, public health, increased the expectation of life to seventy and beyond. The biological sciences improved agriculture. The internal combustion engine and the oil industry would have been impossible without chemistry. Radio and electronics certainly arose out of science, and the aeroplane would be impossible without science. So it is perhaps legitimate to identify science-based technology as a very remarkable explosion in human artefacts, which is still going on. We may now be at the edge of extraordinary changes in the biosphere as a result of molecular biology and the possibility of creating new forms of life. Rocketry, again emerging from chemistry and physics, has opened up the possibility at least of space colonies and the extension of human beings to the solar system.

Any great expansion of any class of species, such as science-based technology represents, will have drastic effects on the older species of the ecosystem. It is not surprising that we are seeing widespread extinctions, partly a consequence of the expansion of the human race itself. Here again, this is a result of science-based artefacts and enormously increasing food supplies. On the whole, the tremendous increase in know-how that has resulted from science has economised on both energy and materials. The rise of the electric power industry, for example, goes along with a marked increase in real output per unit of energy input, in spite of the fact that there are substantial energy losses in the production of electricity. Skyscrapers economise on space, cars economise on time, perhaps at some cost in diseconomies of space as we adapt our cities to what have been described as large four-wheel bugs with detachable brains. Aeroplanes economise on time, perhaps at the cost of energy, but certainly compared with sailing ships. The expansion of the human race threatens wildlife with which human beings compete for space and, perhaps, for shelter and food. Improved techniques in exploiting resources, whether they be oil and gas, coal deposits, or even the whale and deep sea fisheries, can easily result in over-exploitation, even turning what is potentially an inexhaustible resource into an exhaustible and finally an extinct one.

Then there is the problem of pollution. Processes of production, virtually without exception, produce both 'goods' and 'bads'. The tremendous expansion of the production of goods has also resulted in a great expansion in the production of bads—pollution of the atmosphere, rivers, lakes, and now the oceans. Up to a point an ecosystem seems to be able to absorb the pollutants fairly easily, perhaps because they open up new niches for things that will use them. Even oil spills encouraged oil-using bacteria, and a certain amount of industrial excrement can be recycled. Nevertheless, there does come a point when pressure on the ecosystem forces sudden collapse into a much simpler and more primitive set of species. Ecosystems seem to be remarkably resilient to changing parameters, but beyond a certain point there is a drastic transformation, like the transformation from forest to tundra arising from a minute change in average annual temperature. The rise of science-based technology is certainly worrying in this regard. There is a possibility of increasing strain on the total world ecosystem to the point where very radical transformations can by no means be ruled out.

The most obvious catastrophe, of course, would be major nuclear war, which, according to some estimates, might set evolution back a billion years to grass and cockroaches, eliminating all mammals, reptiles, birds, and trees, as well as humans, through a combination of radioactivity and destruction of the ozone layer. The CO_2 problem, though less dramatic, is causing increasing concern. If we burn up all our fossil fuels, there is very little doubt that the CO_2 proportion of the atmosphere will increase. There seems to be a substantial probability that this will produce a 'greenhouse effect', letting in high-level radiation from the sun but stopping low-level radiation from leaving the earth, so that the earth becomes warmer. There seems to be general agreement that the poles will warm up much more than the equator, which will alter the weather systems of the earth profoundly. Any major melting of the ice caps could produce dramatic changes in sea level and, as a large proportion of the human race lives very close to sea level, this could be an enormous catastrophe. Areas that are now fertile might turn into deserts and areas where the climate would become more favourable are apt to be those northern regions where there is not much soil, thanks to the last ice age. Even though an increase in CO_2 would be generally favourable to plant production, it is at least a plausible hypothesis that for the human race as a whole the results of this ecological change would be highly adverse, although by no means fatal, as nuclear war might be.

Another interesting question is whether the diminution of the total biological gene pool as the result of widespread extinctions of both flora and fauna could lead to an ecological catastrophe. One problem here is that both agricultural productivity and medical advance represent, in a sense, a race between human intervention and adaptive mutation in parts of the ecosystem which are adverse in human terms, such as crop and forest pests, mosquitoes and viruses. Substantial diminution of the total gene pool could land us in a position where there would be no way of counteracting the genetic changes in the adverse species. Offsetting this possibility, of course, is the possibility that greatly increased knowledge of our genetic structures will enable us to re-create species. It is certainly not impossible that we might even re-create the dinosaurs, so human know-how could easily offset the genetic impoverishment that is now going on and even lead to an enormous expansion of genetic richness.

There is certainly legitimate anxiety about the development of a society in the world ecosystem which is increasingly dependent on the exhaustion of exhaustible resources, and even of potentially renewable ones. Oil and gas will certainly be much harder to find in fifty years, coal in perhaps three hundred years. The breeder reactor in some form may give us electricity for thousands of years, but electricity is a very imperfect substitute for fossil fuels, simply because it is so hard to store. It is not inconceivable that we might be able to solve this problem, but the problem must be difficult or we would have solved it long ago. Solar energy is abundant in total but very diffuse, and again raises an acute problem of storage. The sun does not shine at night.

The using up of concentrated materials in mines may be even more serious than the problems of energy. Here there is no through-put on which to fall back, as in the case of energy from the sun. If we have energy, we can recycle. But recycling is always imperfect and there is a strong tendency for human activity, particularly for science-based technology, to dissipate concentrated materials and diffuse them around the earth in innumerable dumps, or even in the sea. The apocalyptic fear of collapse when we run out of everything cannot simply be brushed aside. On the other hand, the continual increase in human knowledge and know-how keeps postponing the evil day when everything will be gone. How long it will continue to do so, we really do not know.

What 'environmentalism' really amounts to is the application of human valuations to the total state of the planet. The fact that we live in complex ecosystems does not mean that we are helpless. Just as agriculture distorts the ecosystem of a field in directions which are favourable to human valuations, producing grain rather than useless weeds and brambles, so we can think of social policy as a kind of social agriculture, diminishing social species that are harmful by human valuations and expanding those which are beneficial. We still have a long way to go before we develop the institutions that can guarantee social policy of this kind. A great deal of present social policy, particularly that concerned with national defence, is potentially catastrophic. Nevertheless, the human learning process goes on; it is very hard to wise down after you have wised up. We cannot go back to Eden after eating that apple, and only more knowledge can get us out of the jams in which the application of inadequate knowledge may land us. The solutions to the problems of technology may by no means always mean a higher technology, but they always involve more knowledge.

3. The machine: icon of economic growth

Peter Mathias

Economic historians, just like enthusiastic contemporaries—tourists and commentators—who observed the emergence of the new high-productivity technology of iron machinery and steam power in the eighteenth century, have been mesmerised by the machine—the icon of industrial man, at once the symbol, the measure and the means of our collective modernisation. The Industrial Revolution, from the beginning of its historiography, became the story of key inventions in the cotton industry (Kay's flying shuttle, Hargreave's spinning jenny, Arkwright's water frame, Crompton's mule, Richard Robert's automatic mule and the power loom), of smelting iron with mineral fuel, puddling and rolling, and, above all, of steam. Splendid, marvellous machines, from Newcomen's first commerically working steam engine of 1712, have—rightly—always drawn admiring audiences at the time and dominated the pages of subsequent textbooks. As the 'Corn-Law' rhymster, Ebenezer Elliott, exulted:

> Oh! there is glorious harmony in this
> Tempestuous music of the giant Steam . . .
> Comingling growl and roar and stamp and hiss
> With flame and darkness! like a Cyclopean dream
> It stuns our wondering souls that start and scream
> With joy and terror.[1]

The 'wondering souls' of historians of technology have been stunned ever since. In fact, although regular obeisance was made to these inventions, in almost ritualistic terms, thus acknowledging the central features of innovation in terms of mechanical devices, the economic historians proper—those whose names are most well known, such as Arnold Toynbee, William Cunningham, the Webbs and the Hammonds—did not give much space in their books to actually describing and discussing the productive consequences of the technical changes brought by the great inventions which they all said were fundamental. Arnold Toynbee's *Lectures on the Industrial Revolution of the Eighteenth Century in England* set the canon in 1884. He spoke of 'the great mechanical inventions' and summed up the process: 'Passing to manufacturers, we find here the all-prominent fact to be the substitution of the factory for the domestic system, the consequence of the mechanical discoveries of the time.' The spinning jenny, the water frame, the mule and self-acting mule, steam power, the power loom and transport improvements were then identified.[2] But little more than one paragraph is devoted to this theme in the whole book, which is much more concerned with the social consequences of economic change than the growth of the productive system itself. The text gives the impression that growth in productivity is taken for granted; not dismissed, but subsumed within the listing of inventions as so obvious a feature of change that it was not worth space to delineate. Thus was consolidated a tradition of commentary and analysis which took the wealth-making process for granted and concentrated upon problems of the distribution of income, working conditions, oppression by landlords and employers, environmental deterioration, urban squalor and the like.

In their short textbook, *The Outlines of English Industrial History*, William Cunningham and Ellen McArthur followed a similar path. The book devotes very little space to industry with almost nothing on 'modern' industry. Six pages only are concerned with the classical inventions in cotton and steam power.[3] Again, discussion is more concerned with the consequences—particularly the adverse consequences—of the process of industrialisation and urbanisation for the working classes affected by the changes. The employment-creating effects of the expansion of trade associated with industrial growth and railways were identified more specifically than the growth of wealth derived from enhanced productivity.

> It may be said that labourers, generally speaking have not suffered by the introduction of machinery but only one class or another which possessed a kind of highly specialised skill that is superseded by some machine. This is a real loss but it is a limited one which must be set off against the general gain—to the consumers in cheapness and to labourers generally through the subsequent expansion of trade.[4]

The book as a whole has relatively little to say about industry, particularly modern industry, despite its title.[5]

Another popular, and long-lived textbook by G. Townsend Warner followed the same pattern, led by the statement: 'The story of the amazing development of English industry in the eighteenth century is mainly the story of mechanical inventions.'[6] The text does extend the range of inventions to pottery, chemical bleaching and machine-making, with more extended descriptions (though over only twelve pages) and a chapter on the mechanisation of transport.

William Cunningham's major synthesis in three volumes, *The Growth of English Industry and Commerce in Modern Times*,[7] offered less detail than Townsend Warner, being within the main tradition of the economic historians of resonant generalisations with very little explanation, description or analysis. The text of Volume III begins with the ringing assertion: 'The period which opened with Arkwright's mechanical inventions has been the commencement of a new era in Economic History, not only of England but of the whole world.'

It marked 'one of the great stages in the growth of human power to master nature' and he compared its historical significance with the discovery of the new world or the sea route to India. 'England was the pioneer of the application of mechanism to industry', ran the summary, 'and thus became the workshop of the world.'[8] But an extremely small part of the text was then devoted to describing or analysing the productive consequence of the inventions, with a single paragraph on steam power, invoking James Watt.[9]

In short, the main tradition of British economic historiography during the second half of the nineteenth century shows a quite extraordinary disparity between the extent of the claims made in generalisations about the importance of the mechanical inventions—seen as virtually the sole source of impetus for industrialisation—and the failure to explore the process in any detail. It was all so important, so self-evident, that it could be taken for granted. Readers did not need historians to elaborate what was in the consciousness of everyone—or so it seemed.

In fact, the main elaborators of the story of the mechanical inventions, the principal guardians of the myth that their creators alone produced the Industrial Revolution, were not the economic historians (who, for the most part, accepted the same conclusions) but the popul005ers and biographers, culminating in the astonishing *œuvre* of Samuel Smiles. Here the hagiography of the inventors merged with that of successful businessmen—coalescing where the fruits of inventive genius were subsequently harvested as wealth emanating from a business set up to embody the innovations. The inventors/entrepreneurs, as creators of employment, were the saviours of the poor, and were the source of the nation's new-found wealth and power, the originators of Britain's great gains, as compared with the fortunes of other nations, whose chances of emulation depended upon them acquiring the results of British inventive genius and British skills. The tradition became enshrined in a long bibliography with titles such as *'Fortunes made in Business'* and *'The Romance of Industry'*, with emphasis on inspirational literature for the aspiring young.[10] The sophisticated, detailed analyses of the productivity gains derived from mechanisation, such as Charles Babbage's *On the Economy of Machinery and Manufactures* and A. Ure's *The Philosophy of Manufacture*, form only a very small element in the tradition.[11]

The list of Samuel Smiles' main biographies, which sold in astonishing numbers, enshrines the central tradition about the primacy of mechanical invention: *Men of Invention and Industry, Industrial Biography, Lives of the Engineers, The Autobiography of James Nasmyth* (which he edited and ghosted), and several other individual biographies. On the title page of *Industrial Biography* is the invocation from Carlyle: 'The true Epic of our time is not Arms and the Man but Tools and the Man—an infinitely wider kind of Epic'. The essence of the upward path of mankind was identified as the progressive human command over brute nature: wealth and progress, power and civilisation were conjoined through the efforts of the inventor, the industrialist and the businessman. In particular, Smiles' work became an invocation to the engineer, whose life's work was devoted to the furtherance of these aims. As the preface to *Lives of the Engineers* put it, they were 'the principal men by whom the material development of England has been promoted'.[12] This ideology of the engineer also embraced individualism, force of character, hard work, determination, thrift, and Protestantism in an encomium of the bourgeois virtues, expressed in a litany of highly successful homilies such as *Self-Help, Thrift, Industry, Character, Duty, The Huguenots in England and Ireland,* and *Life and Labour*.

In this established tradition, giving primary (if not exclusive) causation for change to inventions and mechanical innovations, later generations of economic historians sought to relate technical progress with economic advance, and usually assumed that technical changes of this nature were the essential instruments of, even if not the originating stimulus for, economic change. And most, in fact, also assumed implicitly that technology was, indeed, the principal initiating force for change, for industrialisation, even if they did not seek to measure its contribution. Thus, for example, one might well draw the inference from many generalisations in the tradition of this historiography, that steam power was providing a major contribution to the motive force of British industry and mining, and central momentum for the economy, by the end of the eighteenth century. In fact, von Tunzelmann's recent measurement of the social savings of the Watt steam engine over the Newcomen engine puts the figure at 0.11 per cent of current national income, and, over water power, at 0.2 per cent of national income.[13]

The precise implications of such a quantification of social savings still present difficulties, in my view. The conclusion is often drawn that (given a rate of real growth of the economy of 2 per cent per annum), social savings of 0.2 per cent mean that, in the absence of such gains, the economy would have taken a further month or so (10 per cent of a year) to have achieved the same output. But measuring gains *ex post facto* in this way, when designating the various sources of economic growth, creates conceptual and methodological difficulties, quite apart from measurement problems, if the results of the exercise are then used to make judgements about the strategic role of technical change in the dynamics of economic growth, *ex ante*. The strategic importance of steam power, and increments to steam power at the margin, is not revealed by the percentage of total horsepower supplied from steam engines. By breaking constraints at the frontiers of technological advance, steam power could enable more traditional forms of power, and other aspects of technical change and economic development to develop. Steam-powered spinning mills in the cotton industry, for example, helped to induce a massive expansion of handicraft technology in the weaving side of the industry in the two generations after 1780. Steam-powered pumps in deep mines overcame similar constraints in the expansion of output, by mainly pick-and-shovel methods in coal mining. Despite such complications, however, it is clear that steam power provided only a very small fraction of the total horsepower which powered the expansion of the economy in the eighteenth century—water, wind, animal and human[14]—and more generally, that the dynamics of economic growth and industrial expansion were very much more complicated than a tradition of explanation confined to mechanical improvements suggested.

Since 1945, economic historians, like development economists, have acknowledged that the dynamics of growth contain much greater complexities. Great emphasis is now placed upon such relationships as agricultural change, the population equilibrium, the dynamics of demand, property rights and legal processes, institutional and political structures, and motivational dynamics (whether those of entrepreneurs, landowners or the labour force). The processes of invention and innovation themselves are argued to depend in large measure upon response to need, expressed in changing relative factor prices. The gap between formal knowledge of techniques and successful innovation, the even wider gap revealed in the dynamics of diffusing new techniques, and the relationship between average and best practice technology at any one time showed how technical change can be, to a degree, a dependent variable in the process of growth—no doubt a necessary, but evidently not a sufficient, condition in itself for advance. The analysis of technology, *ex ante*, that is to say, when studying its role in the dynamics of economic change, appeared in a significantly different perspective than when looking at the results of change *ex post facto*.

Within the process of technological change itself, significantly different emphases have been made which have also slowly altered our conceptions of the nature of innovation. The earlier tradition stressed heroic innovation: technological change was realised through dramatic new inventions by known individuals who became famous for their achievements. The history of technology was a pantheon of great names. This complemented the assumption that the driving force for progress lay in a small group of strategic industries, or leading sectors —for the Industrial Revolution, the cotton industry, iron and steam power above all. More recent studies have revealed the process of innovation to be much more diffused across the economy, much more gradual, evolutionary and anonymous than the heroic tradition assumed.[15] One example is that of steam power itself. Within the same basic technology, the 'efficiency' of Cornish low-pressure beam engines increased almost three-fold between the original Newcomen atmospheric engines of the 1710s and the inventions of James Watt. Watt's discrete heroic inventions of the separate condenser and the 'double acting' mechanism more than doubled the level of efficiency but then, between the 1790s and 1840s, efficiency rose a further four-fold within the basic Watt technology—in response to gradual improvements in technologies of construction and gradually increasing (but still low) pressures. These gradual, cumulative, continuum-style increases in productivity came from 'learning by doing', from improvements by obscure artisans and captains of Cornish mines in quite a different milieu from that of James Watt.[16]

A further paradox then took place. Just when economic historians were becoming more cautious and circumspect in their views about technical change, behold, the machine became deified once more in a quite new academic subject —industrial archaeology, which became one of the fastest growing industries in contemporary Britain. Admiration for the new in technology in contemporary Britain sometimes appears to be outweighed by an overwhelming fascination with surviving instances of traditional technology—the Industrial Revolution in aspic. This degree of technological nostalgia, which no other nation can rival, has produced extraordinary results—a new academic discipline, complete with learned societies, journals, conferences and university courses (but not yet, I think, a university chair). It made a fortune for certain publishers who realised the extent of the public interest in industrial archaeology. New industrial and maritime museums have been springing to life, month by month, even week by week, to cater for weekend, holiday and school visits, where ancient industrial artefacts no less than ancient castles and country houses are now prime targets for the English deciding where to go in their cars.[17] Remote, uneconomic branch railway lines, when finally abandoned by British Rail as hopelessly unprofitable, became crowded out when returned to steam by enthusiastic amateurs and run as a leisure industry. The implications of such a widespread and deep-rooted fascination with technology, in particular the technology of the past, are significant in ways too numerous to investigate here.

While the economic historian has seen this great wave of interest in the history of technology break over his subject from one direction, from a quite different point of view economists have been adding their voice to the debate. Here is another paradox. The mainstream of economic theorising in western countries, archetypal figures such as Marx and Schumpeter apart, had contrived not to bring technical change into a conceptual framework dominated by distributional analysis under conditions of static equilibrium. Once long-term growth and dynamic analysis became a central issue for economic theory after 1945, the sources of productivity increase were seen at once to lie at the heart of the matter. Development economics, after all, had as its primary goal the reduction of the great and growing gap in wealth between rich and poor nations which was now, in a new world of United Nations and international agencies, a main focus of attention. 'With a few exceptions,' wrote Nathan Rosenberg in 1976, 'it is only in the past twenty years or so that economists have attempted to relate the subject [technological innovation] in a systematic way to their analysis of long-term development.'[18]

In retrospect, we can now see that the critical conceptual advance came in 1954–7, when several eminent American economists produced a technique which offered a means of distinguishing the elements in the growth of national wealth which derived from productivity increases, rather than from incremental inputs in resources, labour and capital. Critical articles by Fabricant, Abramovitz, Solow and Denison revealed that almost all the increases in national wealth in the United States in the very long term (from 1870), as well as in more modern periods (1909-49; 1909-29; 1929-57; 1950-62), sprang from productivity increases rather than from the increase in measured inputs per unit of resources at static levels of productivity.[19] Typically, only 10 or 15 per cent of growth could be identified as responses to extra capital and labour when measured at the productivity levels of the beginning of the period. Clearly, productivity, and all that lay behind increments in productivity were the key to economic development. 'Its source', concluded Abramovitz in 1956, 'must be sought principally in the complex of little understood forces which caused productivity, that is, output per unit of utilised resources, to rise.'[20]

For most economic historians, judging such matters by empirical evidence rather than from a conceptual framework, this revelation was a demonstration of their own basic, if instinctive, judgement. But it was to identify a mystery rather than a solution. What were the sources of productivity growth? The technique of analysis which had allowed productivity gains to be measured in comparison with the growth of inputs captured such productivity gains as a residual; and such a concept was the aggregative result of gains from all sources, wheresoever derived, net of the inputs of resources, labour and capital at static levels of productivity. Further attempts to explore the residual and make allocations within it, for example, by measuring the qualitative gains from a more educated labour force by equating salary and wage levels to marginal product, were based upon more challengeable assumptions.

The issue was summed up by Salter in 1960 in acknowledging that:

> The difficulty, however, is that the interpretation of even the simplest measures of productivity raises a host of very complex problems. For behind productivity lie all the dynamic forces of economic life: technical progress, accumulation, enterprise and the institutional pattern of society. These are areas where our understanding remains rudimentary.[21]

More recently, Nathan Rosenberg's conclusions in 1971, when reviewing attempts to 'assign a precise numerical value to represent the contribution of each factor aiding growth', were deeply pessimistic: 'The limitations in the data and the methodological and conceptual difficulties of all measurement procedures render such a goal perhaps forever unattainable.'[22]

By such a circuitous route, the economists have been led towards the traditional stance of the economic historian in relation to technical change. Without quantitative identification, subjective assessments prevail and the technical literature is suffused with assumptions (sometimes identified as such but more commonly unguarded) that productivity increases mean the results of technical change, acknowledged to be associated with investment in human capital, better education and the like, but essentially the pay-off comes through improved technology. 'There is widespread (although by no means universal) agreement', wrote Nathan Rosenberg, 'that this process [technological innovation] is the primary cause of long-term economic development.'[23] Much depends upon the limits we set upon such terms as 'technical' or 'technological' or 'innovation'—where semantic imprecision covers conceptual uncertainty—but there is, at the centre of the debate, the same assumption as that long held, more instinctively, by economic historians, that productivity increases ultimately derive from improved technology in a specifically embodied form—the technology of production.

However, at this point a sleight of hand often occurs which can deceive an innocent eye about the sources of productivity gains—and I would argue that this has been common to economists, economic historians, and governments as well as amongst wider constituencies. Undoubtedly, technological change, growth based upon advancing technology with the new products and new industrial capacity flowing from that technology (not forgetting the improved and/or cheaper existing products which also result from investment in new technology) lie at the epicentre of this process. But the diagnosis of just needing more technical change, with the higher rate of investment to embody this, appears to me, as a general economic historian, to be too narrow a diagnosis, and is in a long historical tradition of being too narrow. The efficient utilisation of investment is at least as important as its aggregate percentage to the national income, while gains in productivity do not come alone from the installation of new technology—but involve also the commitment of workers, the efficiencies of management and organisation, the provision of specialist financial and business professional services, the financing, distribution, selling and market-orientation of products. The productivity of identical plant in virtually identical factories operated by the same multinational companies has been shown to differ greatly when installed in different countries, and even between different firms in the same country. Capital per unit of output is higher in Britain than in some richer industrial countries precisely because of less efficient utilisation and lower output from comparable plant. Before the full potential productivity gains of embodied technology can be captured, a very much wider range of relationships has to be mobilised.

When technological or engineering priorities exclude the commercial, the economy can get lumpy investments it cannot afford or products which cannot sell at a profit. Victorian engineers in Britain produced the most expensive and least technically co-ordinated railway system in the world. Before 1914, the British car industry made products which were over-engineered—too expensive, denied the possibility of long production runs, too good technically to gain a wide market—the virtues of the mechanical engineer triumphing at the expense of the production engineer. And we have had Concorde more recently. Some present-day governments (of both parties) in Britain, as elsewhere, like some economic historians and contemporary tourists, have been too impressed by dramatic instances of the latest technology when making judgements about the sources of productivity. In a similar way, some industries (particularly, at the moment, the car industry), quite apart from being intrinsically important to the economy in terms of size, employment, value-added and spread-effects through subcontracting, become invested with a totem or fetish quality, whereby they symbolise in the public mind the fate of the entire economy and become a test of national viability, success against foreign competition, patriotism and even a sort of collective national virility. It is, therefore, salutary to remember that of the three richest nations of Europe in per capita terms—Sweden, Switzerland and Denmark—none has large-scale aircraft, basic electronics or nuclear power industries; two have no car industries and the third only a highly selective one.

One of the roots of these attitudes, I would argue, derives from the common assumption, inherited from the long historiographical tradition that has been discussed, that productivity is just what happens on the shop floor, in the factory or on the farm—it stops at the factory gate. Hence a high level of investment within the factory or on the farm will by itself produce the results we all seek. We can, therefore, divide the nation into productive workers wearing overalls, hard hats or white coats, and unproductive workers wearing white collars. The other side of the same limiting assumptions about the sources of productivity is the presumption that a market is something given—a fixed autonomous entity—and that those who serve the market, administer organisations and supply services (in contrast to those who create the machines)—the bankers, the brokers, the salesmen and the accountants, the market-research workers, and particularly the advertisers and publicity men (who have double doses of original sin to expiate)—are at best the non-productive wagon-train of the army, at worst the parasites of the system, the drones in the hive. They suffer all the suspicion and hostility which has historically been reserved for the middleman since organised economic life began. Mediaeval hostility to financial and commercial intermediaries of all kinds, expressed in standing laws against forestalling, regrating and engrossing and in the championship of the 'just price' against the unfettered operation of the market, established the tradition in centuries when the market, for many reasons, was in no condition to operate effectively. *Homo sapiens* has been hunter, herder, grower and maker. *Homo fabricator* has an essential function; he who supplies services is an appendage. Production is what counts; distribution just happens. Again, the textbooks of economic history, reflecting priorities in research, concentrate on the productive system, and have had very little to say about distribution and markets.

These atavistic, instinctive assumptions, which always have a propensity to break surface when all is not well with the economic performance of the country, have long historical roots. In fact, since these same early centuries, the objective process of economic evolution has seen the providers of services become steadily more important as necessary collaborators with the producers, whether in farm or workshop. The larger the scale of output from the single plant or locality, the greater the productivity of the machine, the more dependent innovation becomes upon the advance of scientific knowledge; the wider the marketing range, the more extensive the capital, the more complex the organisation of production, then the potential for expansion, together with the possibilities of realising the full technical limits of productivity, are alike dependent upon more effective provision of services which cocoon the productive function on all sides. Such providers of services are part and parcel of the evolving system which has produced greater wealth: a necessary, if not sufficient, condition. In turn, such services were also critically dependent upon new technologies.

The growth of services relative to physical production (whether measured by value of net output or by differential growth in the labour force) has proved to be partly a consequence of modernisation and the growth of wealth. With increases in income beyond certain thresholds, the utility or satisfaction derived from more services increases faster than that derived from more goods. In all advanced societies, the role of government, and the services dispensed by government, has increased relative to the national income. The parameters of sophistication and costs in health, education, entertainment, travel, leisure activities and the like, expand exponentially with income. Some other increments in the demand for services represent countervailing—or self-negating—responses in the advance of professional expertise. Tax accountants and lawyers prosper in symbiosis with the rising numbers of similar professionals employed by the tax authorities. Increasing wealth encourages increasing litigiousness. Examples are infinite. The accounting convention whereby all this countervailing activity is assumed to be a net addition to the national income may be challenged—but the same query would then doubtless be posed over the utility of some products that find a market and a place in the 'value added' of the manufacturing sector.

Despite such doubts, the progressive expansion of services in the richest, most sophisticated economies and societies is also a consequence and integral aspect of evolution of business and industry itself. The mass production of goods utilising the full potential of improving technology implies an incremental development of R&D, design, planning, distribution and marketing services. An increasing scale or complexity of business organisation requires greater financial and managerial controls. Widening markets demand increased investment in transport and communications. Such relative trends towards services are, in fact, greater than the occupational censuses usually reveal because of the growth of white-collar functions within the occupational groups designated as employed in industry. The cumulative, even exponential, productivity of machine industry, which shows no sign of a downturn from any failure of the pace of innovation, brings this as a virtually inevitable consequence. The dawn of the automatic factory simply maximises the trend.

These broad structural trends can be quickly documented. In the richest economies in the world, the proportion of the national income contributed by services (excluding wholesale and retail trade, communications and transport) is now above 30 per cent, and this has been rising rapidly since 1945—in the United States from 26 to 37 per cent, in Sweden from 26 to 38 per cent, in West Germany from 21 to 32 per cent, in France from 20 to 31 per cent, and in the United Kingdom from 25 to 33 per cent. The inclusion of trade, transport and communications would bring the present percentage up to 50 per cent of the national income. In countries of middling wealth, such as Argentina or Brazil, the equivalent percentage contributed to the national income by services is 23 to 25 per cent, and for the poorest range of countries, such as India, it is usually 14 to 15 per cent.[24] Thus the great rise in wealth, led by the advanced nations since the Second World War, as long before, has been associated with this progressive trend towards a services economy and relatively away from agriculture and industry. In the spectrum between the richest and the poorest nations at a single point of time there is also a very close correlation between the level of national wealth per capita and the relative importance of the contribution of services to the total national income. It needs to be said that this process of 'de-industrialisation', if such be its name, follows a long-term evolutionary trend, not being just the product of short-term depression or crisis; nor, as a gradual phenomenon, is it the immediate cause of high unemployment or stagnation (although it may well be associated with slower rates of growth amongst the richest economies).

More specific measurement is possible to document the growth of business services in the British economy during recent decades, a trend which has been little short of dramatic, but has been much overlooked.[25] Since 1954, the most rapidly expanding large occupational group of the total working population has been that of those providing 'financial, business, professional and scientific services'. They more than doubled in strength from 2.26 million in 1954 to 4.77 million in 1977, their share of the working population almost doubling from 9.3 per cent to 18.1 per cent. In contrast, the share of the work-force in manufacturing industry shrank steadily from 37.9 per cent (9.12 million) of the whole to 27.9 per cent (7.35 million), and that of the labour force in construction from 6.2 per cent (1.49 million) to 4.8 per cent (1.27 million). Since 1951, insurance, banking, finance and business services has been consistently the fastest growing sector of the economy in terms of output, save only for gas, electricity and water.

This expansion of services, in fact, has also wrought a quiet revolution in Britain's balance of payments during the past generation, no less than in the structure of the national economy.[26] Net surpluses earned by services have been offsetting deficits in the balance of commodity trade since the eighteenth century, but their relative importance and the contribution of different constituent services have changed markedly in recent times.

Net earnings from shipping, for example, which were previously by far the largest of all the surpluses from services, making up about two-thirds of the whole before 1914, have gone into deficit, while the earnings of financial services have risen most rapidly.[27] The net surplus produced from all invisibles (interest and profits as well as earned income from services) rose from £273 million in 1967 to virtually £2,000 million in 1977, while the visible trade deficit grew from under £600 million to over £1,700 million. The role of services in total exports increased from less than a fifth to over a quarter. Credits from financial services alone increased from just over £100 million in 1964 to over £1,360 million in 1977. Moreover, some earnings from financial services lie outside these totals, being hidden within other categories, such as contributions to the earnings of British firms overseas or repatriated business profits. Swift expansion overseas by British accountancy firms, in company with banks and insurance companies, may be one reason why finance and insurance services contribute a significantly higher percentage to gross domestic product in the United Kingdom (11.7 per cent in 1970) than in either France (8.5 per cent) or Germany (6.4 per cent).

Such are the new dimensions being wrought in the British economy, and in its dealings with the world, by the contributions of services. Laments over the export performance and poor output and productivity record of British industry too often divert attention from the excellent record which British business has enjoyed in supplying commercial, financial and professional services to the world. Overseas success in such diverse fields as international banking, insurance, aviation (if no longer shipping), ship-broking, the international metal exchanges and commodity markets, retailing, merchanting, property companies, construction firms, auctioneering, accountancy, and—dare one add?—British medical, professional and academic expertise, stands as proof of internationally competitive standards in sectors of the economy of progressive importance. Indeed, the freer from restrictions that trade in services is internationally, the more successful have British firms tended to be.

This has been a silent revolution, still largely undocumented by historians. Success does not breed evidence like failure. Quite apart from the predilections which historians, like governments, have had for production, many services have not spawned the documentary data which have provided the staple fare for historians of industry. There have been no armies of oppressed workers in coal mines, satanic mills or the sweated trades to induce elaborate parliamentary investigations, followed by detailed annual reports of official inspectors; no great legislative battles in Parliament to stir the nation and produce publications concerning oppression or crisis; no great strikes to dominate the consciousness of a generation; no sustained threats of nationalisation, or reports from the Monopolies Commission; no extensive record of failure, with lame ducks to be supported or the National Enterprise Board to be wooed. This is not to say that the path of expansion has always been smooth, but, compared with the history of textiles or shipbuilding or mining, it has been relatively trouble-free.

In conclusion, some of the qualifications to be applied to these trends in economic performance and changes in the economic structure and structure of employment need to be stressed. The growth of services in the economy, whether by value added, or exports or employment, is not to be seen as a possible effective response to the short-term, urgent, major problems of very high unemployment, the fall in industrial output and GDP of recent years, and the threat, from bankruptcy, fall in rates of profits, and collapse in investment and research expenditure, to the private sector of industry. The greatest short-term prospects for recovery in rates of growth, employment and exports alike will come from the revival of industrial fortunes, now severely afflicted by short-term depression, maximised by the downturn in activity in the world economy, as well as by longer term structural change.

Direct redeployment from industry to the service sector in the short run on a massive scale is not a feasible option to fuel recovery. Such structural redeployment is a long-run gradual process, dependent upon favourable demand conditions as a whole. In addition, as in previous periods of structural redeployment, those laid off from industry may well not be suited for alternative employment, or willing to take what is offered. Redeployment has to be seen in terms of region as well as role; it also has the rhythm of inter-generational change as new cohorts choose educational options, take decisions about the acquisition of skills, view the various prospects of the job market, and make their choice according to the options.

A flourishing industrial base, particularly in the private sector, is critical for the fortunes of the economy: critical for employment prospects, critical for higher growth rates, critical as the source of wealth from which transfer payments, directly and indirectly, have to be made through government to sustain welfare, education, health, social infrastructure, cultural and all those other expenditures which modern states, whatever their reigning political philosophies, are increasingly called upon to sustain throughout the world. Not least, in present circumstances in the UK, will a private sector revival be critical to sustain failing competitiveness in public sector industries. And when North Sea oil runs out, or declines from its maximum output, at the beginning of the next century, the export gap will have to be closed, in large part at least, by industrial exports and indigenous industrial production taking the place of imports.

But in the longer term, we may suppose that these changes in economic structure will continue, more particularly in the structure of employment to a greater extent than in the structure of output (by value added), because of the fantastic productivity possibilities in manufacturing industry which technology offers in the immediate future. Perhaps the historian, just because he has a professional commitment to such long-term perspectives in the past, can dare to forecast them in the future.

Notes and references

1. Elliott, E. (1958), *Steam at Sheffield*, cited in Warburg, J. (ed.), *The Industrial Mix*, London, Oxford University Press, p. 19. See also Briggs, A. (1982), *The Power of Steam*, London, Michael Joseph.

2. Toynbee, A. (1908), *Lectures on the Industrial Revolution of Eighteenth Century England*, London, Longmans Green, pp. 7, 64-74.
3. Cunningham, W. and E. McArthur (1895), *Outlines of English Industrial History*, Cambridge, Cambridge University Press, pp. 220-7.
4. Ibid., p. 229.
5. The chapter headings tell the story: 'Immigrants', 'Physical Conditions', 'Manors', 'Towns', 'Beginnings of National Economic Life', 'Food Supply', 'Industrial Life' (principally about labourers, poor relief, settlement, prices and wage controls, aliens, personal freedoms), 'Commercial Developments', 'Economic Policy', 'Money, Credit and Finance', 'Agriculture', 'Labour and Capital', 'Results of Increased Commercial Intercourse'.
6. Warner, G. Townsend (1899), *Landmarks in English Industrial History*, London, Blackie and Son, p. 225.
7. Cunningham, W. (1903), *The Growth of English Industry and Commerce in Modern Times*, Cambridge, Cambridge University Press, Vol. III.
8. Ibid., p. 609.
9. Ibid., pp. 626-7.
10. *Fortunes Made in Business* (1884-87), London, Sampson Low & Co., 3 vols. Alfred Marshall's copy is heavily annotated, with close connections with some of the text in his *Industry and Trade* (1919), London, Macmillan. Other typical examples of this genre are Dodd's *Curiosities of Industry* (1859), London, Routledge; Martin, J. (1894), *Chats on Invention*, London; Bridges, T. and H. Hessele Tiltman (1928), *Kings of Commerce*, London, G. G. Harrap.
11. Babbage, C. (1832), *On the Economy of Machinery and Manufactures*, London, Charles Knight; Ure, A. (1835), *The Philosophy of Manufacture*, London, Charles Knight.
12. Smiles, S. (1861), *Lives of the Engineers*, London, p. iii.
13. Cf. von Tunzelmann, G. (1978), *Steam Power and British Industrialisation to 1860*, London, Oxford University Press, pp. 149, 156-7.
14. For example, total horsepower in Watt engines is reckoned at 12,400 in 1800; water power at 50,000-100,000 h.p. Animal horsepower remains unquantifiable. Von Tunzelmann believes that windmills may have provided as much horsepower as all steam engines in 1800 (op. cit., pp. 123, 148, 151).
15. Schmookler, J. (1966), *Invention and Economic Growth*, Cambridge, Mass., Harvard University Press; Schmookler, J. (1962), 'Economic sources of inventive activity', *Journal of Economic History*, 22 (1): 1-20; Parker, W. N. (1972), 'Technology, resources and economic change in the West' in Youngson, A. (ed.), *Economic Development in the Long Run*, London, Allen and Unwin; Rosenberg, N. (1976), *Perspectives on Technology*, Cambridge, Cambridge University Press.
16. Mathias, P. (1979), *The Transformation of England*, London, Methuen, Chs 2 and 3.
17. See, for example, Hudson, K. (1980), *Good Museums Guide*, London.
18. Rosenberg, N. (1976), *Perspectives on Technology*, Cambridge, Cambridge University Press, p. 61; see also Jorberg, L. and N. Rosenberg (1982), *Technical Change, Employment and Investment*, Department of Economic History, Lund University, Sweden, esp. pp. 8-27.
19. Abramovitz, M. (1956), 'Resource and output trends in the United States since 1870', *American Economic Review*, 40 (2); 5-23; Fabricant, S. (1954), 'Economic progress and economic change', *Annual Report of the National Bureau of Economic Research*, New York; Solow, R. (1957), 'Technical change and the aggregate production function', *Review of Economics and Statistics*, 39 (3): 312-20; Denison, E. (1967), *Why Growth Rates Differ*, Washington, Brookings Institution. The field is admirably covered in Rosenberg, N. (ed.) (1971), *The Economics of Technological Change*, Harmondsworth, Penguin.
20. Abramovitz, M. (1956), 'Resource and output trends in the United States since 1870', *American Economic Review*, 40 (2): 6.
21. Salter, W. (1960), *Productivity and Technical Change*, Cambridge, Cambridge University Press, p. 1.

22. Rosenberg, N. (ed.), (1971), *The Economics of Technological Change*, Harmondsworth, Penguin, p. 317.
23. Rosenberg, N. (1976), *Perspectives on Technology*, Cambridge, Cambridge University Press, p. 61.
24. All data from *United Nations Statistical Yearbooks*. 'Other Services' include 'financing, insurance, real estate and business services, community social and personal services, public administration and defence'. The individual categories are not separately shown.
25. See data in Blackaby, F. (ed.) (1978), *De-Industrialisation*, London, Heinemann.
26. Sargent, J. (1978), 'UK performance in services' in Blackaby, F. (ed) *De-Industrialisation*, London, Heinemann.
27. Insurance, banking, commodity trading, merchanting of other goods, brokerage, legal earnings—and accountancy earnings.

4. Technology beyond machines
Stuart Macdonald

Introduction

The sort of games psychiatrists play would reveal the close association of technology with machines which is widely perceived but which is not evident in official dictionary definitions of either word. The *Concise Oxford Dictionary* makes no mention of technology under machine, nor of machine in its rather cryptic definition of technology as the 'science of the industrial arts'. Yet it is unusual to imagine technology without machines—not necessarily huge iron monsters crushing objects in factories, but machines in the wider sense of the inanimate shaped to perform useful tasks. Perhaps the main distinction between science and technology is that science is allowed to be abstract while technology is pre-eminently practical. Notions of technology divorced from physical means of application are difficult and uncomfortable. Similarly, machines—in the broad sense—are seen as technology, the means by which science is reduced to practice. The more modern and intricate the machine, the more it is technology and the less a fortuitous assembly of basic elements. Thus, while a capstan is definitely technology, a numerically-controlled capstan is much more so. The height of high technology industry is measured chiefly by the sophistication of the machinery it produces; the proximity to science of product, rather than of process or application, is apparently critical. The discipline of economics has been outstanding in its confident confusion of technology with machines: technology has traditionally been subsumed within capital—within the 'buildings and machinery' category, in fact—and technological change measured by changes in capital input. Consequently, a new computer has been assumed to have the same effect on production as a new factory shed of similar cost. This chapter is concerned with that part of technology beyond the machine, a part whose importance remains largely unappreciated.

Machines in perspective

It is, of course, quite possible for technology to involve no machines at all. For example, a crop rotation is technology, and new crop rotations are decidedly technological change; indeed, that very change in technology was probably more significant than any other in agricultural improvement in Europe during the eighteenth and nineteenth centuries.[1] Similarly, it is quite possible to view change in administrative or managerial practices as technological change once perceived associations with machinery are removed.[2] Technology can be regarded simply as the way things are done, and technological change as the adoption of what are thought to be better ways of doing things—a definition not too dissimilar to that provided by Schumpeter long ago.[3] Obviously machines are likely to be involved in this process, but it is very difficult to envisage technology which is completely embodied in a machine. Even total automation requires technology beyond its hardware; the hardware is useless without software (and what is sometimes called 'peopleware') to direct and control its operation, and the whole lot is pointless without organisation of raw materials entering the factory, and of finished products leaving the factory.

If, then, machines do not fully embrace technology, it is obviously necessary to establish an understanding of just what does. Technology is really the sum of knowledge—of received information—which allows things to be done, a role which frequently requires the use of machines, and the information they incorporate, but conceivably may not. Such a definition is not as catholic as it may at first appear; it is reserved for practical achievement and excludes aims, ideals and philosophies. However, the definition extends far beyond techniques, which are really just the tools of technology and, as Nelson has put it, are bereft of the 'logy'—the theory—part of technology.[4] The history of regional input-output analysis has provided a fine example of the distinction between technique and technology, and of the inadequacy of the former when applied without the corpus of knowledge encompassed by the latter.[5] Ironically, a better example still may be the use of technique to model and forecast the impact of technological change itself, In such cases, mastery of the technique and ignorance of the technology—the technology of technological change, as it happens—has sometimes produced results that are of neither practical nor intellectual value.[6] Table 4.1, for instance, is the product of a single government programme using input-output technique to model the impact of technological change on Australian industries in 1990/91. Clearly, there have been some desperate recalculations between 1978 and 1981. The programme saw technological change in terms of labour productivity. That was forecast by projecting historic trends and by incorporating the predictions of a handful of experts, experts apparently preoccupied with technical promise rather than commercial reality. As one report confessed: 'The forecasts of most experts were relatively insensitive to a fairly wide range of price variations'.[7]

Technological change, then, may be defined as the addition of new knowledge to old knowledge, usually to allow things to be done in what are thought to be better ways, and sometimes to do new things altogether. Technology being inherently practical, experience is supposed to establish whether the new ways actually are superior to the old—or even to none at all. A major problem in such assessment is that it is much easier to gauge the performance of that part

Table 4.1. Persons required per million dollars of annual output by selected Australian industries in 1990/91 (1971/72 prices)

Industry	1978 forecast for 1990/91*[a]	1979 forecast for 1990/91[b]	1981 forecast for 1990/91[c]
01.01 Sheep	63	44.8	49.4
01.02 Cereal grains	21	29.8	46.1
01.03 Meat cattle	15	22.4	32.3
01.04 Milk cattle and pigs	68	108.3	65.2
01.05 Poultry	26	41.6	36.4
03.00 Forestry and logging	60	61.6	53.4
04.00 Fishing, trapping and hunting	89	89.5	73.8
23.02 Man-made fibres, yarns etc.	20	36.7	32.9
23.03 Cotton, silk, flax yarns etc.	43	50.3	57.9
23.06 Textile floor covering	8	25.7	13.1
24.01 Knitting mills	36	38.6	44.3
24.03 Footwear	24	53.1	37.1
33.01 Scientific equipment etc.	9	10.2	54.7
36.02 Gas	22	22.1	25.4
37.01 Water, sewerage and drainage	18	18.6	32.0
48.03 Other repairs	64	65.6	45.7
52.01 Rail transport	73	75.4	50.9
53.01 Water transport	–	8.2	23.5

* Preliminary and exploratory figures.
Source: a. Chapman, D., *Forecasting Technological Coefficients*, Bureau of Industry Economics Working Paper No. 1, Canberra, 1978, pp. 57-9.
b. Bureau of Industry Economics, *An Analysis of the Economic Implications of Alternative Technological and Social Scenarios for the Year 1990-91*, Submission to Committee of Inquiry into Technological Change in Australia, Canberra, 1979, Part 1, Appendix 3.1.
c. Bureau of Industry Economics, *The Long-run Impact of Technological Changes on the Structure of Australian Industry to 1990/91*, Research Report No. 7, AGPS, Canberra, 1981, pp. 18-20.

of the technology closely associated with the machine; the technology of the rest of the process is more nebulous and much harder to assess, which may result in the simplistic assumption that if the machine is technically efficient, the technology in which it operates is likely to be efficient too. That new machines are indicative of efficient technological systems and old ones of inefficient systems has been disproved in many developing countries.[8]

The linear innovation process

Clearly, if technological change requires new knowledge, the process by which information is created for technological change is worth consideration. The initial adoption of the new is innovation, and the traditional model of technological innovation is of a process inaugurated by research, and involving the trauma of development before the climax of innovation. Such a model assumes a convenient linearity which is almost certainly unjustified, but which is also assumed in much government science and technology policy, presumably for the same convenience.[9] Consequently, the main element in the process is judged to be research, and policy to stimulate research is justified on the grounds that it leads to innovation. Revelations that development is much more expensive, time-consuming and uncertain than research have done little to qualify enthusiasm for what is imagined to be the seminal part of the innovation process.[10] What appears after innovation is seen as another process altogether—that of diffusion. Geographers are especially fond of plotting rates of diffusion over space and time, and their approach typifies the assumption that innovation is the end of one process and diffusion the start of another, as if innovation marked the point by which technological crystals had formed and were ready for immutable distribution. Until fairly recently it was fashionable to measure technology lag in terms of the time taken for an innovation to spread from its origin,[11] implying that an innovation can be regarded simply as a technological module requiring only to be plugged in to other organisations, other economies and other cultures.

Yet technology is the totality of information which allows things to be done, and total information is unlikely to arrive in a crystalised package from the conventional research and development process. All that can reasonably be expected to emerge from that process is information which must be supplemented by other information before things will be done. Some of this other information will be embodied in the hardware of associated technologies if what is termed 'technological compatibility' is to be achieved. Babbage's computing engine failed because the technology of mechanics was less advanced than that of mathematics,[12] but current telecommunications networks provide examples aplenty of the problems of technological incompatibility.[13] Some of the additional information is analogous to the software necessary to render computer hardware operational, but much of it is information required to effect complementary organisational change. For example, a point-of-sale computer system adopted by a supermarket cannot work at all without software information to supplement the hardware information embodied in the machine, but its usefulness remains limited to fairly basic check-out functions unless there is organisational change to allow the supermarket to take advantage of, say, the new range of management information made available. Clearly this is part of a post-innovation process—though it may provide the basis for further innovation—rather than part of the traditional innovation process, but equally clearly it cannot be entirely separated from that traditional innovation process. Innovation is an integral part of what is in reality a much larger and more complicated innovative process.

The innovative process

A major advantage of the use of a model venturing well beyond innovation is that it loosens the perceived connection between technology and machinery.

The existing quasi-linear model, extending from pure basic research to innovation, becomes increasingly machine-centred as it progresses. But beyond innovation, machine orientation diminishes rapidly and technological change is revealed as a broad information process of which the encapsulation of information in machinery can be only a part. Currently, R & D statistics are obsessed with the creation of technological information in the secondary sector. Although R & D activity in the tertiary sector is acknowledged, it fits uneasily with the preconception that technology is machine-orientated. Thus, for example, the Australian government gives industrial R & D grants for work on computer hardware, but only exceptionally for work on computer software.[14] Of the seven activities distinguished in the definitive Frascati Manual as part of the scientific and technological innovation process—R & D, new product marketing, patent work, final product or design engineering, tooling and industrial engineering, manufacturing start-up, and financial and organisational changes—all, even the last, are seen as tributary to hardware.[15] There is little attempt to disaggregate R & D statistics in the tertiary sector; in Australia the business enterprise category with the second largest R & D expenditure is the 'other not elsewhere classified' category of the 'other industries' section, a virtually anonymous section responsible for about a third of Australian industrial R & D performance.[16] Though the tertiary sector in developed economies is now responsible for most employment and wealth creation, and is the location of most information activities, it is still not acceptable to consider it a major source of technological information—presumably not because it does not create this information, but because it does so without also creating machines. Such a distinction is not constructive and is discouraged by the notion of an innovative, rather than an innovation, process.

Participants in a process

This new concept might also affect the kudos afforded the participants in technological change. At present, high status is associated with basic scientific research and status declines as R & D becomes more applied, involving engineers rather than scientists. After innovation, status plummets and that afforded the salesman is scarcely worth measuring. Yet the characteristically British tradition that a salesman need know nothing about the product, that if a man can sell one thing he can sell anything, contrasts with philosophy elsewhere which insists that a salesman must know the product well if he is to channel information both to and from the market.[17] The Texas Instruments procedure of encouraging scientists to accompany major developments to the market would be anathema to those who espouse the British tradition, and yet it is merely practical acknowledgement of the existence of a single information process extending well beyond innovation.[18]

Inasmuch as technological change is viewed as a process of information flow as well as information creation, that flow is imagined to be a simple process involving pushing promising research results into industry for development. Even that, though, may not be easy when research has been conducted in organisations distant in every way from industry. Better formal channels, such as on-line computer networks, to facilitate this flow may prove counter-productive if they are constructed by demolishing the irregular and informal channels along which much technological information seems to flow.[19] Informal, personal contact is likely to be discouraged when individuals are not only in different sorts of organisations, but also perceive themselves to be performing discrete functions—with appropriate status—in a process that ceases with innovation.[20] A key element of success in new high-technology firms is the working relationship between the technical expert and the commerical expert; both play essential information roles in the innovative process, but the co-ordination of information embodied in individuals with very different characteristics is not easily achieved.[21] Unless there is some success in co-ordination, scientists will continue to talk mainly with other scientists, engineers will talk mainly with other engineers, and neither will see much profit in wasting time with salesmen.

Beyond innovation

Although a staunch defence of the linear model of innovation—invention leading to innovation, basic research to applied research and then to development, science producing technology—would be rare, it is very convenient when dealing with such an uncertain process as technological change to assume that everything hangs from research. As Gannicott has pointed out, even when the importance of 'reverse linearity' would suggest that government industrial R & D incentives have actually been counter-productive, convenience is a more fundamental goal of policy than efficiency.[22] The linear model permits justification of policy incentives for research and permits use of measured resource input to research as an indicator of output. While there probably is some linearity in the innovation process, it is far from clear, even in that limited process, just where the impetus starts and which direction it takes. In the full innovative process, the origin and direction of information flow are even less certain. Invention, it seems, may spring from innovation, development from marketing information, applied research from production problems.[23] Thus, consideration of an innovative, rather than an innovation, process makes much less tenable the assumption, discernible in much literature on patents, that invention is the key element which automatically initiates a chain reaction.[24] Much of the usefulness of new technology depends on the way it is used, on information derived from, and applied by, the users themselves. The information supplied by even nineteenth-century agricultural labourers seems to have been critical to the usefulness of the new machinery they operated.[25] So crucial is this user information reckoned to be that a process of horizontal innovation has been postulated—by the doyen of traditional diffusion studies, as it happens—in which users themselves are largely responsible for the creation and dissemination of technological information.[26]

The extension of the process required for technological change beyond innovation means, of course, that much discussion of the effects of technological change acquires new parameters. Technological change ceases to have an impact on, say, employment or productivity because change in such areas is itself regarded as part of the innovative process. Thus organisational change affecting employment or productivity is part of greater organisational change needed to complement the product of innovation to create a total technological information package. Unemployment should not be seen as a potential and unfortunate consequence of technological change, but as a possible ingredient of the total package of change as fundamental as any change in machinery. The problem of technological determinism is really a problem caused by the assumption that new information associated with machinery is superior to other information required for technological change. Associated with that supposed superiority is the elevated status of those responsible for creating information to make machinery rather than to make machinery useful. Because the technocrat does not recognise the importance, and perhaps even the existence, of other information than that which he possesses, he naturally assumes that only technical solutions are relevant to technological problems. General reluctance to challenge this assumption has suffered those most expert in the technical information of technology to expound on other parts of the innovative process with which they have little familiarity, and in which they are often ignorant of their lack of information.[27] For example, in 1975 the Chairman of the Australian Atomic Energy Commission declared:

> Our technological civilisation produces a continuing stream of problems of a most complex technical character. In many cases judgement must be made on issues which cannot be proved absolutely one way or the other. Only a small proportion of the population is capable of understanding issues of this sort, even if they were to make the effort. Many elected representatives, though not all, are in the same situation. The experts must in the end be trusted.[28]

The scientist's contribution to science policy often provides evidence of this situation: in 1977, one of Australia's leading research chemists, and then Chairman of the country's major science policy advisory body, blithely ignored the endeavours of generations of historians:

> [The] social effects [of the stirrup] over the years were enormous. It led to the development of a specialised corps of mounted soldiers who needed considerable support not only from foot-soldiers, but also from other men to feed and care for the horses. The mounted soldiers took to wearing armour and special methods had to be used to get them into the saddle. It was these

mounted soldiers who became the squires and aristocrats and the medieval society was born—through the stirrup.[29]

Further, exponential growth in the quantity of information involved in the innovative process has increased and intensified specialisation[30]—most evident in the exclusive ritual language of many experts[31]—and has helped to isolate each speciality from the rest of the process. Peer group assessment within existing disciplines struggling with new masses of information has discouraged serious inter-disciplinary work and the formation of new disciplines. Thus, the study of technological change, for example, is fragmented, with little concern for the total process.[32] Vocational education has also discouraged information flow among even existing disciplines, and instead of alleviating specific manpower problems inherent in the process of technological change, has probably exacerbated them by sacrificing the flexibility that is a necessary accompaniment to all change.[33]

Information and the innovative process

Perhaps the greatest penalty for associating technology with machines is exacted through the use of information technology itself. The peculiar characteristics of information as a good—for example, that information remains with the seller even after the buyer has taken possession, and that the buyer cannot be allowed to know what the information is before purchase—make information difficult to classify alongside other economic goods, and may be partly responsible for its neglect by economists.[34] Yet the importance of information in a developed economy is manifest, and much new technology, especially that associated with microelectronics, computing and telecommunications, is designed to cater for information. New machinery offers extraordinary improvements in ability to assemble, store and process information. As might have been predicted, though, plummeting hardware costs have been accompanied by soaring software costs, reflecting the growing problem of providing machinery with information about what it should do with information. There is a much more serious problem. Information is still a most unfamiliar good, and it is understandable that machines which are able to produce cheaply more information should be welcomed as warmly as machines to produce cheaply more of any other good. Over-supply, apparently, is not possible—the more information the better:

> Computer-based management information systems have facilitated more effective management techniques by improving the extent and availability of information on which decisions are based. It is axiomatic that the wider the range of relevant information available to management, and the more accurate and timely that information, the better is management able to monitor and evaluate organisational progress and development.[35]

What has been ignored in the euphoria surrounding the declining production cost of information is the consumption cost of the information produced.[36] Resources are required to use information, and the more information produced, the more resources are required. An obvious example would be the time taken to read, digest and decide what to do with the information on a computer printout. More subtle would be the cost of making a better decision when half a dozen alternatives are increased to several thousand, or when sheer quantity of information exceeds even short-term retention capacity and blurs the distinction between relevant and irrelevant.[37] New machines for the production of information have made more information available for use, but have not necessarily made more information more useful. Thus they satisfy the definition of technological change only in that what has been adopted is thought to be a better way of doing things. It can be thought to be so because the costs of change—which are largely information costs—are often not assessed (particularly if they are internal to an organisation), and are anyway not considered to be part of technological change. Again, it is the perception of technology as machine-orientated that is mainly responsible for this situation. If technology and technological change are ever to be better understood, a new and grander perception altogether is required, a perception of an innovative process which extends beyond the limits of innovation, far from machines, to encompass the entire environment of technological change.

Notes and references

1. See Macdonald, S. (1980), 'Agricultural response to a changing market during the Napoleonic Wars', *Economic History Review*, 2nd Series, **33** (1): 59-71.
2. See, for example, Brown, R. (1981), 'You have done research—whether or not you know it', *Government Accountants Journal*, **30** (2): 46-69; Teece, D. (1980), 'The diffusion of an administrative innovation', *Management Science*, **26** (5): 464-70; Alderson, W. (1965), *Dynamic Marketing Behavior*, Homewood, Illinois, Richard D. Irwin, pp. 259-79.
3. Schumpeter described technological change as 'any "doing things differently" in the realm of economic life'. Schumpeter, J. (1939), *Business Cycles*, New York, McGraw-Hill, Vol. 1, p. 84.
4. Nelson, R. (1982), 'The role of knowledge in R & D efficiency', *Quarterly Journal of Economics*, **97** (3): 453-70.
5. Jensen, R. and S. Macdonald (1982), 'Technique and technology in regional input-output', *Annals of Regional Science*, **16** (2): 27-45.
6. See Mandeville, T., S. Macdonald, and D. Lamberton (1980), 'The fortune-teller's new clothes. A critical appraisal of IMPACT's technological change projections to 1990/91', *Search*, **11** (1/2): 14-17; Chapman, D. (1980), 'Why telling fortunes is better than telling fairy tales: a reply to a Grimm view of SNAPSHOT', *Search*, **11** (6): 179-82; Mandeville, T., S. Macdonald and D. Lamberton (1980), 'The wisdom within the fairy tale—towards a happy ending?', *Search*, **11** (6): 183. See also Macdonald, S. (1982), 'Technological change and structural adjustment' in Heys, K. and B. Langworthy (eds), *ECTA Readings in Economics*, Sydney, Economics and Commercial Teachers' Association of NSW, Vol. 3, pp. 62-9.
7. Bureau of Industry Economics (1981), *The Long-run Impact of Technological Changes*

on the *Structure of Australian Industry to 1990/91*, Research Report No.7, Canberra, AGPS, p. 114. fn. 18.
8. See, for example, Robinson, A. (ed.) (1979), *Appropriate Technologies for Third World Development*, London, Macmillan; Bhalla, A. (1979), *Towards Global Action for Appropriate Technology*, Oxford, Pergamon.
9. See Macdonald, S. (1982), *The Individual Inventor in Australia*, report to the Industrial Property Advisory Committee; Macdonald, S. (1983), 'Faith, hope and disparity. An example of the public justification of public research', *Search*, 13 (11/12): 280-8.
10. See Scherer, F. (1980), *Industrial Market Structure and Economic Performance*, Chicago, Rand McNally, 2nd edn., p. 410. See also Mandeville, T., D. Lamberton and E. Bishop (1982), *Economic Effects of the Australian Patent System*, Canberra, AGPS, pp. 24-41.
11. For example, Tilton, J. (1971), *International Diffusion of Technology. The Case of Semiconductors*, Washington D.C., Brookings Institution.
12. See Hollingdale, S. and G. Toothill (1975), *Electronic Computers*, London, Penguin, pp. 15-62.
13. Macdonald, S., T. Mandeville and D. Lamberton (1981), 'Telecommunications in the Pacific Region—impact of a new regime', *Telecommunications Policy*, 5 (4): 243-50. The slow adoption of semiconductor electronics by the car industry is partly explained by incompatible technologies. See Braun, E. and S. Macdonald (1982), *Revolution in Miniature. The History and Impact of Semiconductor Electronics*, Cambridge, Cambridge University Press, pp. 200-2.
14. Australian Industrial Research and Development Incentives Board (1982), *Annual Report 1980-81*, Canberra, AGPS, p. 1.
15. OECD (1981), *The Measurement of Scientific and Technical Activities (Frascati Manual)*, Paris, pp. 15-16.
16. Australian Bureau of Statistics (1982), *Research and Experimental Development, All Sector Summary, Australia 1978-79 (8112.0)*, Canberra, p. 9.
17. Aldcroft, D. (1975), 'Investment in and utilisation of manpower: Great Britain and her rivals, 1870-1914' in Ratcliffe, B. (ed.), *Great Britain and her World*, Manchester, Manchester University Press, pp. 287-308.
18. Braun, E. and S. Macdonald (1982), *Revolution in Miniature. The History and Impact of Semiconductor Electronics*, Cambridge, Cambridge University Press, p. 141.
19. Macdonald, S. and T. Mandeville (1982), 'Sources of technological information in Australia', paper delivered to 52nd Australian and New Zealand Association for the Advancement of Science Congress, Sydney.
20. Macdonald, S. (1980), 'Personal communication in research and development' in Callebaut, W. *et al.* (eds), *Theory of Knowledge and Science Policy*, Belgium, University of Ghent, pp. 255-69.
21. Macdonald, S. (1979), 'The need to succeed', *Journal of General Management*, 4 (3): 74-83. See also Mansfield, E. (1981), 'How economists see R & D', *Harvard Business Review*, 59 (6): 98-106; Schumpeter, J. (1939), *Business Cycles*, New York, McGraw-Hill, Vol. 1, pp. 84-6; Sawyer, G. (1978), 'Innovation in organisations', *Long Range Planning*, 11 (6): 53-7.
22. Gannicott, K. (1980), 'Simple economics and difficult policies: the case of public money for research and development', paper delivered to 50th Australian and New Zealand Association for the Advancement of Science Congress, Adelaide.
23. See Gibbons, M. and R. Johnston (1974), 'The roles of science in technological innovation', *Research Policy*, 3: 220-42.
24. Macdonald, S. (1981), 'Patents in perspective' in *The Economic Implications of Patents in Australia*, Canberra, Australian Patents Office, pp. 21-38.
25. Macdonald, S. (forthcoming), 'Agricultural improvement and the neglected labourer', *Agricultural History Review*.
26. Leonard-Barton, D. and E. Rogers (1981), *Horizontal Diffusion of Innovations: an Alternative Paradigm to the Classical Diffusion Model*, Working Paper No. 1214, Sloan School of Management, M.I.T.; cf. Rogers E. (1967), *Bibliography on the Diffusion of*

Innovations, Diffusion of Innovations Research Report No. 6, Department of Communication, Michigan State University.
27. Macdonald, S. (1981), 'Technological change and the expert' in Ward, W. and M. Bryden (eds), *Public Information. Your Right to Know*, Brisbane, Royal Society of Queensland, pp. 53-9.
28. Baxter, P. (1975), 'Some comments on Ann Mozley Moyal's "The Australian Atomic Energy Commission: a case study in Australian science and government"', *Search*, 6: 458.
29. Badger, G. (1977), 'ASTEC: planning for science and technology in Australia', public lecture, Griffith University, Brisbane, 10 September.
30. See de Solla Price, D. (1969), 'Policies for science?', *Melbourne Journal of Politics*, 2: 61-8.
31. Reinecke, I. (1982), *Micro Invaders*, Ringwood, Australia, Penguin, pp. 8-10.
32. See Nelson, R. and S. Winter (1977), 'In search of a useful theory of innovation', *Research Policy*, 6: 36-76.
33. See Macdonald, S. (1979), 'A matter of quality', *General Engineering Transactions*, Institution of Engineers, Australia, 3: 49-53; Macdonald, S. (1978), 'Education of engineers in Britain', Institution of Engineers, Australia, *Conference on Engineering Education*, National Conference Publication 78/6, pp. 17-22; Macdonald, S. (1979), 'Unemployment among qualified scientists and engineers in Australia', *Search*, 10 (6): 224-7.
34. See Lamberton, D. (ed.) (1971), *Economics of Information and Knowledge*, Harmondsworth, Penguin.
35. Elliott, R. (1981), 'Technological advances and the Australian banking system' in Goldsworthy, A. (ed.), *Technological Change—Impact of Information Technology*, Canberra, AGPS, p. 175.
36. Lamberton, D., S. Macdonald and T. Mandeville (1982), 'Productivity and technological change: towards an alternative to the Myers' hypothesis', *Canberra Bulletin of Public Administration*, 9 (2): 23-30.
37. See Klapp, O. (1982), 'Meaning lag in the information society', *Journal of Communication*, 32 (2): 56-66.

PART II
THE ECONOMIC THEORIST'S DILEMMA

5. Can we explain technical change?
Arnold Heertje

It is doubtful that we shall ever be able to 'explain' technical progress to our own satisfaction since technical progress almost by definition involves the appearance of the unforeseen.[1]

This chapter is concerned with the possibility of an explanation of technical change. The problem will be discussed from the restricted view of an economist who realises that deep questions of a philosophical nature are involved. However, the temptation to emphasise these philosophical issues too much will be avoided by the partial substitution of common sense.

The structure of the chapter is as follows. First, what might be understood by explanation of technical change is described in some detail. Thereafter, a section is devoted to the relationship between technical change and the production function. The conclusions are then illustrated by means of two recent publications.

Explanation of technical change

Although the process of technical change is a composition which is difficult to unravel, for the present purpose three components must be distinguished at a theoretical level: the development of new technical possibilities, the application of new methods of production and of new products, and the diffusion of both technical knowledge and its applications.[2]

Although most of the literature refers to the production of technical knowledge as technical change, the essential core of such change is the development of new technical possibilities. This broader interpretation is preferred because the range of technical possibilities can increase without, as the literature mandates, an increase in the stock of technical knowledge, for example, by education or the discovery of a new metal with unique and special properties. Further, the application of new technical possibilities is often included in the description of technical development, and this interpretation is, in fact, more commonly used in everyday language. There is a connection between the application of techniques and the creation of new technical possibilities if such an application leads to new technical knowledge. Both the application of new methods of production and the introduction of new products will be treated as separate features of technical change. Diffusion can be thought of as the spread of new technical insight and as the distribution of its applications through space and over time.

Can the development of new technical possibilities be explained? The answer is entirely dependent upon the type of explanation sought. If the explanation is satisfactory that a new method of production is, in the end, the result of creativity combined with the use of scarce resources, the answer would be positive. But ability to explain this development would be less certain if what were sought were the causes of creativity, the role it plays, and its relation to the allocation of scarce resources. Further, if the criterion for a successful explanation were that predictions are possible on the basis of the causes indicated, the answer would be negative—even if the predictions were not necessarily true. The exact specifications of new methods of production or of new products that will become available in the future cannot be predicted. What can be predicted is that the stock of technical knowledge of a certain kind will be higher in coming years if it is decided to allocate resources in that direction with the aim of producing new technical knowledge. Casting the problem in terms of cause and effect, it may be possible to give a description of the relevant causes, but not to establish the relationship between these and the exact nature of new technology. On the other hand, it is possible to establish a relationship between new technology in general and the allocation of resources on both the theoretical and practical levels. Even a nation's technology policy is based on global explanation in this sense.

Whether we can explain the application of new technical possibilities is less problematic because it is known, more or less, what will be applied. Of course, it is still uncertain whether the new production methods and products will be successful, but at least their technical appearance is known. The problem then is to explain which new methods and products will, in fact, be applied. This may still be a complicated problem, as not only economic, but also social, psychological, sociological, and cultural factors play a role. Economic theory, however, is in principle capable of systematically explaining why certain technologies are applied while others are not.

The same kind of reasoning may be applied to the problem of the diffusion process; again, it is not an easy job to explain the exact pattern of diffusion, of new insights, and of the applications, but there are no insurmountable obstacles.

The remarks made so far suggest that it is useful to address the crucial question in more detail: can the development of new technology be explained? In order to answer this more accurately, the concept of the production function must be discussed, and particularly whether a shift in the production function can be equated with technical development.

Technical change and the production function

The production function portrays all the efficient production techniques existing at a given moment in time. These techniques form a mathematical set of combinations of the means of production that ensures the maximum output of a certain product. There is a single maximum output of a certain product. There is a single maximum output belonging to a certain combination of the means of production. The production function is the expression which translates the quantities of the means of production into the quantity of output produced. Empirical production functions, which link the actual recorded output with the actual factor inputs, should be sharply distinguished from theoretical productions, which describe the connection between production and the factors of production *ex ante*. In the latter the production function tells us what technical possibilities are available to the firm. It is useful to make a sharper distinction than is customary between an increase in the range of technical possibilities and an increase in technical knowledge; this is what underlay the distinction made between technical development in the narrow sense and that in the broad sense. A firm acquires new technical knowledge from elsewhere, so that technical possibilities but not technical knowledge increase in an objective sense. The production function reflects the technical possibilities, which do not depend on technical knowledge and its diffusion alone. The description of technical possibilities by the production function shows a certain diversity. Different assumptions are made about the extent of substitution between the factors of production, and diminishing, constant and increasing returns to scale can in principle be taken into consideration. In some cases the production functions reflect increasing, diminishing and constant returns in different intervals, while in other cases they do not. The fact that production is a process in time can be expressed.

It is tempting to examine the significance of these characteristics of the production process, but such an attempt is hampered by the conceptual difficulty that the production function, regarded as an *ex ante* description of the technical possibilities of the enterprise, cannot be observed directly. The actual choice of certain combinations of means of production depends, for example, on the objectives of the firm and the prices of the means of production. The changes observed in the use of the means of production reflect a shift in the equilibrium position, and only in very exceptional cases do they give an idea of some points along the production function. It is admittedly possible to discover directly what pattern the relationship between production and the factors of production follows in the course of time, but this is not an observation of the production function in the sense that is intended here. In this situation discussions of the empirical characteristics of the production function are somewhat speculative. Economic theory copes with this situation by formulating production functions that are as general as possible, that is, production functions that can accommodate several variants. The *a priori* exclusion of certain possibilities often condemns empirical investigations to failure, which emphasises the value of generalised production functions.

Apart from the construction of general production functions, in which no comment is made *a priori* about a number of aspects, there is another approach in economic theory. Production functions indicate the maximum possible output for each combination of the means of production, and therefore rely on a technical maximisation process. For this reason, they are not derived from the technical possibilities, but are instead postulated. The characteristics of the production function are, however, not independent of the character of the set of all the technical possibilities. This character can be expressed by means of a set of postulates, and leads to the development of a production theory based on the theory of sets.

Such a production theory, which gives a complete description of all the technical possibilities, has the added advantage of providing an elegant way of accounting for some other aspects of the production technology, such as the joint production of several different goods. Besides, it gives a description of the technical possibilities that is not based on aggregation.

A question arises as to whether the production function reflects the state of technology available to the firm in a given period of time. This cannot be taken for granted. The set of all the technical possibilities from which the production function is derived depends not only on the state of technology in the sense of technical knowledge available to the enterprise; the possible combinations of the means of production and the nature and volume of the corresponding output depend also, for example, on the quality of the available factors of production —labour and capital—and the extent to which division of labour is possible within the firm. Education, the diffusion of technical knowledge, employee living conditions, environmental hygiene, laws of all kinds, and the size of the firms are thus involved in the production set. In short, therefore, the technical possibilities portrayed by the production function depend not only on the available technical knowledge, but also on all the circumstances that affect the potential combinations of the means of production in a given period.

If the state of technology is construed in a broad sense, so that all the technical possibilities are encompassed by it, then it is fully expressed by the production function; if, on the other hand, it is construed in the narrow sense, as meaning available technical knowledge, then it can no longer be said that the production function reflects the state of technology exclusively. The function reflects all the technical possibilities determined not only by technical knowledge but also by the other conditions of production.

One might conclude that the production function is a purely technical relationship, since all the conditions mentioned converge in the set of today's efficient and technically feasible combinations. In the framework of a static model, therefore, the production function can be regarded as an exogenously given relationship from which other economic characteristics can be explained.

However, a production function also has a past: today's production function is partly the result of a series of past decisions, which almost always relate to the use of scarce resources that can be used as alternatives. To this extent it is the product of past economic decisions. In the framework of a dynamic analysis, one must explain today's production function partly in terms of previous economic decisions; the significance of this for economic theory is that the production function is at least partly an endogenous relation.

Economics should abandon the view that the production function is a relation given in an exclusively exogenous manner. As the expansion of technical knowledge and technical possibilities becomes more dependent on the use of scarce resources, the endogenous character of the production function increases, so that economists should consider it one of their tasks to explain the production function on the basis of economic parameters. A corollary of this is that, in the formulation of economic policies, technical development should no longer be treated as a process that is, in principle, not amenable to control.

Relating the present-day production function to decisions in the past by referring to (a) the resources allocated to the development or production of technical knowledge and—broader—technical possibilities, (b) the application of new methods of production and the use of new products, and (c) the diffusion of both development and application in the past, leads to the explanation of today's technical possibilities. All three aspects of technical change play a role in the establishment of the set of the present technical possibilities. Of course, it is still a complicated task to trace the exact structure of the dependence of the new technical possibilities on the earlier allocative decisions, but in principle it can be done.

The complications arise mainly from two sources. First, there is no one-way avenue from the allocation of resources to new technical possibilities because in each period of time the new technical possibilities also determine the quantity and quality of resources that are available for the production of new technical knowledge. In general, the relationship between technical change and growth is a circular one, and there is a danger of explaining technical change by technical change. The second source of complexity involves the building blocks that together make up the act of creativity that is always involved in the process of technological change over time—the brainwave, the birth of the idea, the decisive step so simple after it has been taken, the relevant question, the glimpse of brilliance, the experiment in the research laboratory with unforeseen but essential side-effects. It is well known that the present technical possibilities rely heavily on the inventions made in the past. They are in some way or another part of the explanation we provide when establishing in a systematic way the relationship between the relevant decisions and events of the past and the set of technical possibilities of today. Insofar as it is impossible to pinpoint the role of the creative aspect any further, it remains a black spot in our explanation of technical change. On the other hand, we know that creativity, especially in the technical field, can be stimulated by incentives and by the urgency of the need for a breakthrough. In a rather formal way one could argue that creative labour is also part of labour as a resource. But even then we cannot avoid the impression that it is too ambitious to establish a direct link between the emergence of a technical breakthrough and the nature and significance of the incentives or necessity for the invention.

The degree of circularity in this reasoning is related to the deeper question of whether the spelling out of a causality over time is causality at all. One could argue that the reasoning so far is not much more than the presentation of a historical description over time. Today's effects are often tomorrow's causes, perhaps even by definition.

Whether one accepts the historical description of how technical change comes about as an explanation in terms of cause and effect is to a certain extent a matter of taste and semantics. It seems to depend on the question of whether insight or prediction is the decisive criterion. Even here, there is room for different interpretations. Must it be just insight or systematic insight, and what then is the difference between this and feasibility to predict? And even if we take the strong position that we can speak only of a sophisticated or acceptable explanation—meaning that the theories we produce enable us to predict the future exactly—we often face complications of lack of proof that reality and theoretical prediction coincide, so that a decision cannot be made.

On the assumption that one would attach some weight to the predictive nature of the reasoning, it seems appropriate now to relate the present-day production function to technical possibilities in the future. Production functions have not only a past, but also a future. For the firm, the production function of tomorrow generally has a greater scope than the present one. Can the appearance of a new production function be identified with technical development? If the latter is taken to mean nothing more than the increase of technical knowledge, then the answer must simply be 'no', for a new production function can also be due to an improvement in the quality of labour as a result of education and training. If, however, technical development is taken to mean increase and change in the technical possibilities, then by definition a new production function must be identified with technical development.

Various reasons have encouraged us to distinguish between a broad and a narrow meaning of the concept of technical development. Technical development in the broad and narrow sense behaves to some extent as effect and cause, but it should be borne in mind that there are also other causes of technical development in the broad sense, such as changes in the living conditions of the labour force, improvement of the environment and changes in the size of the firm. Furthermore, technical possibilities affect the production of new technical knowledge.

On the other hand, does technical development lead to a new production function? The answer must be 'yes', both for the broad and the narrow interpretation of technical development. It is widely held that technical change can be regarded as a shift in the production function. Here the starting point is that the new production function arises from the old as a result of one or more of the coefficients being changed. It is not impossible, according to either interpretation of technical change, that a new production function can be derived from the old one by altering one or more of its parameters while keeping its form intact. Technical development in the narrow sense should be thought of as an increase in technical knowledge that causes not a qualitative but a quantitative change in the relationship between production and its factors. Technical development in the broad sense can also be such that the form of the production function does not change with time, only its coefficients do.

Technical development in the twentieth century shows that technical knowledge often undergoes fundamental changes at the level of the firm. Essential changes come about in the production set through the creation of new means of production, new products and new relationships between the goods. These changes have an effect on the production function derived from the production set. New production functions arise which differ from the previous ones in a discontinuous manner. The custom of aggregating capital goods and various types of labour in the construction of the microeconomic production function may have led to the misunderstanding that technical development can be generally characterised as a smooth continuous process. Even when the broad interpretation of technical change is used—and thus attention is focused on new technical possibilities—it must be concluded that new production functions arise on account of radical changes in the social structure, as a result of which the nature and number of technical possibilities for the firm undergo fundamental changes.

Not only is technical change interpreted in economic literature as a shift in the production function, but, conversely, a shift in the production function is often taken as a sign of technical change. What economists quite often have in mind is a simple shift in the production function, with a larger output from the same combination of means of production in a subsequent period of time. A causal connection is then assured between the production functions for different periods. These economists generally assume technical change in the narrow sense. It can now be seen that a shift in the production function can indeed be due to technical change in the narrow sense, but we must reject a simple connection between such a shift and technical change. This is partly because the production function may be shifted in the same way by factors that lie outside technical development in the narrow sense (so that technical development in the broad sense is then the relevant concept), and partly because this approach does not take into account the diversity of the causes and effects of technical development. Even when the shift in the production function is interpreted so broadly that it encompasses any change in one or more of the parameters of the function, the view of technical development still remains restricted by the formulation of the production function, which is the starting point of the argument under review. The production of entirely new goods, and the expansion of technical possibilities by the construction of entirely new means of production, lie outside the scope of this argument. It is evident that an incomplete picture is given of technical development by expressing it as a shift in the production function, for an increase in technical knowledge and the creation of new technical possibilities cannot generally be squeezed into a shifting production function, and the diversity of technical development is better expressed by identifying the latter with the appearance of new production functions.

It is interesting to compare this conclusion with the frequently voiced, and apparently contradictory, view that technical development can be thought of as a gradual and continuous process in time. This view is not really contradictory because in the first place the actual course of technical development depends not only on the expansion of technical possibilities described by the production function, but also on the choice the entrepreneurs make from the wealth of potentially useful techniques. The gradual application of new technical possibilities should not be confused with a gradually shifting production function. But even if the actual cause of technical development is considered to be gradual and continuous, the notion of an invariant causal connection between the production functions for different periods must still be rejected. The gradual creation of new products and new means of production does not detract from the qualitative character of the change, which is not described adequately by a shifting production function, since this only accommodates quantitative changes.

The above argument can be further elucidated by examining the concept of the new production function. An objection could be raised to this argument that the difference between the old and the new production function is not sufficiently sharp, especially when the shift in production function is considered in such a broad sense that it encompasses an arbitrary change in one or more of the coefficients. If the production function is formulated in a sufficiently general form, then the new set of coefficients is always in a certain relationship to the old set, while the form of the function remains the same. In this light, the transition from a Cobb-Douglas production function to a CES (constant elasticity of substitution) function must be regarded as the appearance of a new production function, unless, for example, a sufficiently general VES (variable elasticity of substitution) function is postulated from the outset. A change in the elasticity of substitution is then a quantitative change in a parameter, as a result of which we are faced with a shift in the general production function. Thus each new manifestation is, as it were, immanent in the previous one, provided that this has been described in sufficiently general terms.

Having arrived at this point, we can now profitably return to the set-theoretical production theory from which the production function has been derived. The set of all the technical possibilities in each period is accounted for in a detailed manner in this theory, and—what is particularly important—all the goods are mentioned separately. Each activity is characterised by a quantitative relationship between production and the means of production. The characteristics of the production function that can be derived from the production set are partly determined by the postulates introduced. The meaning of a shift in the production function based on the production set is invariably that the efficient activities undergo a quantitative change. We are then concerned with changes in the parameters which are either accommodated by the design of the old production function, or can be derived indirectly from it by reformulating it in the light of the new production function. In this sense, the new production can always be derived from the old one, and one can speak of a shift in the production function.

We have to realise, however, that technical development comprises, in particular, the creation of new products and new means of production. The production set is then considerably enlarged by the introduction of entirely new processes, and so an entirely new production set is created. The production function derived from the new production set is essentially different from the old one, since it comprises new products and new means of production. Thus technical development manifests itself in the appearance of new production functions that cannot be derived from the previous one, because the underlying production sets differ in nature and number of goods. This shows the great value of the set-theoretical production theory for the analysis of technical development.

The investigation can now turn to the new production sets and their properties. It is not impossible that this will reveal forms of technical development characterised by a temporal consecutiveness that is more complicated than has so far been imagined by economists. The difficulty is that in so far as the new production sets are due to scarce resources becoming available, what new products will be created and what new capital goods will be necessary for production still cannot be predicted. Additionally, inventions that renew the production sets will also be made independently of the economic process. Given all this, the production set and its construction should be the object, rather than the starting point, of analysis in economic theory. Although, in the framework of a static analysis, production sets can be regarded as technical data, the change of one set into the other in the course of time is so strongly influenced by economic factors that the dynamic analysis of technical development cannot ignore the economic effects on the formation of new production sets.

In any case, the analysis of the creation of new technical possibilities should be separated from the analysis of the entrepreneur's choice of optimal production technique, in which profits and cost calculations definitely play a part. Even though it has been assumed that the increase in technical knowledge is to a considerable extent determined endogenously, and thus demands an economic explanation, this is so different from the economic analysis of the choice of optimal production technique that it is desirable to keep the two problems separate.

Is it possible to derive the new production function from a theory about the allocation of resources, in which creative labour plays an important role? In other words, can we predict what the new production function will look like? The answer has to be negative: it is impossible to foresee the exact nature of the new technology. In this sense, we cannot explain the new production function, and therefore technical change. Neither from a theoretical nor from a policy point of view is it satisfactory to leave the matter here.

However, the allocation of resources with regard to technical change is not just a matter of pure speculation. At this stage it seems appropriate, in view of the uncertainty surrounding the new technology, to distinguish between deterministic and stochastic allocation of resources. It is not known how much has to be invested in order to produce new technical possibilities. It is not known how many creative people are available, nor who, in fact, will do the job. But it is known that there is a certain future probability that new technical possibilities in a certain field will emerge if a certain amount of money is invested and creative men and material are combined. Though it is not possible to predict the exact nature of the technical breakthrough, after a certain time something new will be forthcoming. So, in practice, the procedure is to allocate resources in a stochastic way in order to produce the new technology, and to invest more intensely we resort to new technology quickly.

Can technical change, in the sense of predicting new technology, be explained? The answer must be negative if what is meant by new technology is the exact nature of the new products and new methods of production. It is interesting to note that every theory aimed at explanation and prediction and cast into a falsifiable proposition will be falsified by experience, and therefore meets Popper's criterion for proper empirical theorising. The answer must be positive if what is meant by new technology is just that new technical possibilities become available. Admittedly, the latter interpretation is very vague; therefore, our answer must be—on balance—that technical change cannot be explained. However, this conclusion should not lead to the idea that it is useless to study the economic aspects of technical change. On the contrary, if we have to content ourselves with description, we derive insights that can also be applied and that would allow the formulation of a policy with respect to technical change. It remains an important task for economic science to deal with technology and technical change. The production function's endogenous character illustrates that it is to be treated as something that needs analysis, at least in terms of description. It should not be regarded as offering an explanation, by treating it as a starting point for economic reasoning. After the creation has played its role, there are still important features to be explained, such as the diffusion process and the actual choice of new technologies to be applied from the set of known technical possibilities.

Two recent publications

These findings may be illustrated by referring to two recent publications in the field of technical change. The first is a book on the diffusion of innovation by Lawrence Brown.[3] The main emphasis in Brown's book is on the diffusion of new products and new methods of production; innovation is taken in the Schumpeterian spirit of the application of new technology. This implies that the technology considered is known, although the diffusion still may lead to unforeseen side-effects and even new ideas. However, the explanation of the diffusion process in space and over time is a possibility not hampered by the obstacles encountered earlier. Brown, nevertheless, seems to avoid in general the word 'explanation' when spelling out his theory of the diffusion process and shows a certain preference for the word 'understanding'.[4] He presents his book as an attempt 'towards developing a contemporary understanding of the innovation diffusion'.

Brown's theory is a major contribution in the field. It is based on the idea that not only demand factors, such as resistance to adoption of new technologies, but also supply factors govern the diffusion process. Thus, diffusion of innovation is no longer simply a consumer behaviour phenomenon. It is, instead, a much broader topic requiring consideration of institutional behaviour by public and private entities which affect the individual's or household's access to the innovation. On this basis Brown develops four perspectives in his book. The adoption perspective views the diffusion as the outcome of a learning process of the consumers. The market and infrastructure perspective considers the ways by which innovations, and the conditions for adoption, are made available to consumers. The continual technological improvements or the continuity of innovations is covered by the so-called economic history perspective of the diffusion process. The fourth view of the innovation diffusion process is the development perspective. It concerns the impact of diffusion on individual welfare on the one hand and the influence of the level of development on diffusion on the other. Brown's conclusion is that all four perspectives must be considered 'in coming to understand and use the innovation diffusion process'.[5] His book illustrates the main point of this chapter: the fact that we cannot explain technical change in the sense of predicting the character and timing of new technology on the basis of its causes does not exclude the possibilities of contribution to the understanding of essential aspects of the process of technical change.

The second recent publication I would like to refer to in this connection is the book *Market Structure and Innovation* by Kamien and Schwartz.[6] This book is devoted to the mutual relationships between market structure and technical change. Many important questions are dealt with, such as: What is the socially efficient allocation of a given size research budget? What is the socially optimal research budget? Does the market equilibrium yield an efficient allocation of resources to inventive activity? These and other questions are related to what is called the Schumpeterian hypothesis, the idea that the presence of monopoly power and the possibility of realising monopoly profits stimulate technical change. The authors illustrate that from their so-called decision theoretic approach, it follows that firms that succeed in the race to innovate and realise a monopoly profit have less of an incentive to innovate in the next round than those who lost and are therefore earning normal profits. This makes technical change self-sustainable in a market environment. Technical uncertainty regarding the development cost function is taken care of by assuming that the probability of successful completion of innovation at any time is an increasing function of cumulative effective effort at that time. The model also takes into account market uncertainty, which refers to the firm not knowing exactly how much money rivals are spending on development of a substitute innovation. In the same theoretic approach to the relationship between market structure and innovation, the essential assumption in order to describe the race to innovate with several active participants is that no one can be assured of being first.

The analysis of Kamien and Schwartz is very fine and subtle and it leads to many interesting insights about the interaction between market structure and technical change. However, it does not provide us with an explanation of technical change. In fact, Kamien and Schwartz also seem to avoid terminology that refers to explanation, as may be seen from their closing passage:

> Regardless of which approach is employed, static or dynamic, deterministic or stochastic, decision theoretic or game theoretic, optimizing or behavioral, the quest is for a more complete understanding of the economics of technical advance.[7]

Conclusion

Economic theory cannot explain technical change. As the unforeseen cannot be explained, it would be surprising if any science could do the job. We will never be able to predict the exact nature of technical change.

It does not follow that nothing at all can be said about technical change. Technical knowledge, being the product of a production process in which scarce resources are allocated, can be produced. We do not know exactly what will be produced, but we are certain that we will know more after an unknown period. The new technical possibilities can be partly explained in terms of the allocation of scarce resources. In this sense, technical change seems to be a composite of endogenous and exogenous components. If the new technologies have taken a certain shape, the principle obstacle to explaining all kinds of related phenomena seems to be removed. In particular, the diffusion process and the choice out of the set of technical possibilities may be explained in economic terms. Still, the interaction between technical change and other social and economic developments demands much of our innovative activity as economists.

Notes and references

1. Dewey, D. (1965), *Modern Capital Theory*, New York, Columbia University Press, p. 140.
2. See also Heertje, A. (1977), *Economics and Technical Change*, London, Weidenfeld and Nicolson.
3. Brown, L. (1981). *Innovation Diffusion: A New Perspective*, London, Methuen.
4. Ibid., p. 281.
5. Ibid., p. 285.
6. Kamien, M. and N. Schwartz (1982), *Market Structure and Innovation*, Cambridge, Cambridge University Press.
7. Ibid., p. 223.

6. A conceptual framework for modelling the role of technological change
Duncan Ironmonger

Economists have been prone to setting out the conceptual framework for their models of economic systems by starting with the assumption that the technologies used by the system are fixed. Given the many complexities of economic systems, this has been a convenient simplifying assumption. It has allowed many of the operations of, and interactions between, the sectors of economic systems to be studied. However, the assumption of fixed technology is patently at variance with even casual observation. Technological change may be so important that for the economic systems to be properly understood we need to re-examine the basic concepts of our economic models. The focus of attention of economists is obviously conditioned by the structures of the models they use.

Modelling technology

The need for a new framework can perhaps be better understood by considering two contrasting models of the economic system: (i) a model with a fixed technology, and (ii) a model which allows technology to change. The structures of these models in relation to consumption, production and trade would be along the following lines.

(i) *A fixed technology model*

In the consumption sector of this model there would be a given number of commodities, the qualities or characteristics of these commodities would be fixed and these characteristics (the technology of consumption) would be known to all consumers. The focus of attention for economists would be on explaining how changes in prices and incomes alter consumption patterns.

In the production sector of the model of an economy with a fixed technology there would be a given number of industries. The products of these industries would have fixed characteristics, the processes for producing commodities (the technology of production) would not be altered, and the producers or entrepreneurs would all know the available processes of production. The focus of attention for economists would be on explaining how changes in factor prices (costs) and in scale of production lead to changes in demand for labour and capital, and hence to changes in factor shares and income distribution.

A fixed technology model of world trade would have a fixed number of countries with given industries producing given commodities by unchanging processes. There would be a fixed process of trade and transport between countries, and the attention of economists would be on explaining how changes in country costs and scales of production and demand would alter patterns of international trade.

There is thus quite a lot that could be explained about the functioning of economic systems within the conceptual framework of a fixed technology model. However, a changing technology model shifts the focus of attention for the economist.

(ii) *A changing technology model*

In the consumption sector of this model there would be an increasing number of commodities, and the qualities or characteristics of these commodities would be changing through time. Consumers would be engaged in the process of finding out about the technology matrix (the coefficients connecting characteristics to consumer wants), and would be in the process of altering their consumption patterns as they found out about this matrix. Consumers would also respond to changes in prices and incomes, but the focus of attention for economists would be on how consumers obtain product information, on the rates of adoption of new commodities, and on the abandonment of outmoded products.

In the production sector of the changing technology model, products have different characteristics from one period to the next as new products are made, and new processes for production are invented. Producers would be busy finding out about these new products and new processes, and would adopt them (innovate) as advantages were discovered and understood. The attention of economists in this model would be on explaining the rate of adoption of new processes and products, the effect on productivity in industries, and the change in patterns of demand for intermediate products and factors as innovation takes place.

A changing technology model of world trade would have a growing number of countries engaged in international trade in an increasing number and variety of commodities with a changing spectrum of characteristics. The process of trade and transport involved would also be changing its characteristics through time, leading to changes in the technology of international trade. The focus of economists would be on understanding the flow of technological information between countries, on comparative rates of introduction of new commodities in consumption and of new processes in production, and on the changed flows of international trade following from these developments.

The nature of technological improvement

The focus of attention on prices and incomes in the fixed technology model is obviously narrow and may have obscured the nature of the improvements in the world's economic systems brought about through improved technology. Our results may well be biased. Technological change may well be much more than the customary 'residual'.

Evolution in the technique of consumption, production and trade has been the process by which the standard of living has been improved. Perhaps nearly all of the improvements in productivity should be attributed to improvements in technology and technique, only a small part being attributed to the more intensive use of existing 'natural' resources. The total set of resources available as inputs to the economic systems of the world consists of 'man-made' resources, 'man-modified' resources and 'natural' resources.

Man-made resources include all those techno-structures created by man, including tools, machines, buildings, and roads as well as ideas and social structures at every level from the individual and the family through to business organisations and public institutions. Man-modified resources include such things as agricultural areas, pastures, exploited natural resources and harbours. Natural resources consist of the remaining elements of the natural environment that could be used as economic inputs, both biological (such as fish, wild animals, and natural forests), and physical-chemical (such as radiation from the sun, the atmosphere and the oceans).[1]

Man-made and man-modified resources are often regarded as 'technology'. Hence changes in technology would be taken to mean modifications in the attributes of any of these resources. An improved variety of a crop or animal would be an improvement in technology, but so also would be an improvement in the skills or knowledge of the members of a particular craft or occupation. Changes in educational levels leading to improved human performance are just as much an improvement in technology as improvements in the yield of grain from wheat crops, milk from cows or wool from sheep.

In modelling the role of technological change it may be useful to recognise the distinction between 'hardware' and 'software' made in relation to computers and computer programs. Consumers and producers need to know not only the technical characteristics of the commodities they are buying (the food, clothing and household goods or the trained labour, equipment and buildings); they also need to know the techniques for combining these commodities (the recipes of the cookery book or the schedules of the production process). Particular recipes and schedules available to households or business managements may be better or worse, more or less efficient, just as we know computer programs can be efficient or inefficient. Hence, innovation in the software or management techniques of households and businesses may be as important in improving productivity as changes in the characteristics of equipment, raw materials or finished goods.

Diffusion processes

The outline of the changing technology model on page 51 indicates that there are three important diffusion processes to be studied. Each of these involves a process of spreading information amongst consumers, producers and countries, and consequent adoption of changed techniques of consumption, production and trade. They are:

(a) the diffusion within an economy of information about the characteristics of new consumer commodities and the adoption of changed consumption technologies;
(b) the diffusion within an economy of information about the characteristics of new production processes and the adoption of changed production technologies;
(c) the diffusion between countries of information about the characteristics of new commodities and production processes, and the adoption of a changed technology of international trade.

Each of these diffusion processes has at least two stages: (i) the diffusion of the information about new products and processes amongst consumers, producers and traders, and (ii) the diffusion of the actual adoption by purchase or use of the new product or process.

The speed of the process of the diffusion of information relating to the characteristics of commodities and techniques is crucial in the changing technology model. Information also has a role in the fixed technology model, but in that model the information required by the economic agents is mainly about prices and costs, which can change. But the rate of diffusion of information about prices and costs is assumed to be more or less instantaneous and, as technology is assumed to be fixed, no new information is required about technology.

There is a probability that the 'parameters' governing the second stage, the adoption stage, of the diffusion process may be more complex than has been realised. The apparent slowness of the rates of adoption may be due not only to delays in the spread of information, but also to the formation of expectations by consumers and producers about further changes in the characteristics of products or processes.[2] Decisions to postpone the adoption of an innovation can be based on well-founded expectations concerning the rate at which further improvements will become available.

Beyond this, the economic systems themselves are part of technology. These systems include not only the consumption, production and trading sectors, but also institutions such as central banks, monetary authorities, finance ministries, and budget management offices set up to manage the performance of the economic systems. Thus a changing technology model would need to describe an additional process for the spreading of information amongst countries. This fourth diffusion process is:

(d) the diffusion between countries of information about the characteristics of new processes of economic management, and the adoption of changed economic management technologies.

Studies of diffusion processes

Some stages and some elements of diffusion processes have been researched. When the consumption sector of an economy is studied closely, it soon becomes clear that the most significant changes in consumption patterns are related to the introduction of new commodities and the falling away of outmoded commodities.[3] The diffusion of information about new commodities leads to the incorporation of 'efficient' commodities in consumer budgets. The speed at which these successful commodities are adopted has been studied and some understanding of the factors governing success and speed has been obtained.[4] Similarly, the process of innovation in production within countries has been studied and comparative studies of technological innovation taking place in different countries have been conducted.[5] These studies have determined the parameters of the diffusion process in specific cases. The main result seems to be that, for the major products and processes in consumption and in production, innovation is a process which takes many years, sometimes decades.[6]

It should also be noted that the process of technological change in consumer goods leads to further rounds of technological change in the production process as new methods are found to produce new consumer products more efficiently. Consumption innovation opens up possibilities of production innovation. In turn these innovations lead to possibilities of innovation in the process of international trade.

What seems to be missing is the incorporation of these findings into the macroeconomic models of countries and into the world trade and world economy models that are currently in use or in development. These models seem to derive from the fixed technology model outlined earlier. The parameters of the diffusion processes of new products and techniques have not been incorporated into these models. In some cases, technological change has been taken into account in economy-wide and world-wide models through the inclusion of 'residual' trends in productivity, often different for different industries. These residuals are obviously strongly related to the rates of introduction of new commodities and of new techniques for producing commodities.

Even so, the residuals approach does not seem to specify the parameters of the diffusion processes in a logically correct way. The conditions controlling the rates of adoption remain unspecified. Perhaps it is asking far too much to suggest that the economic models should contain all the parameters governing the rates of diffusion of new consumer goods and new producer goods. On the other hand, though, the residuals approach seems bound to lead to poor predictions.

The need for technological change

The economic crisis of 1982 with a renewed surge in unemployment, higher inflation in many countries and intense strain on the international financial system, suggests that there is a crying need for improvements in the control systems for maintaining stability and growth in the economic systems of the world. Improvements in our technology for economic management are urgently required. The advantages to be gained from further improvements in the detailed production and consumption sectors can easily be lost through a failure to establish effective systems for maintaining full employment and price stability. This means we need to promote invention and innovation in the public sector, the sector where these controls are exercised. The breakdown of economic stability is perhaps due to the continued application of outmoded ideas for economic management which have tended to produce inappropriate positive feedbacks. What we need is the discovery and adoption of good ideas for economic control through a process of useful negative feedbacks.

Notes and references

1. Gallopin, G. (1981), *Planning Methods and the Human Environment*, Paris, UNESCO.
2. Rosenberg, N. (1976), 'On technological expectations', *Economic Journal,* 86: 523-35.
3. Ironmonger, D. (1972), *New Commodities and Consumer Behaviour*, Cambridge, Cambridge University Press.
4. For example, Bain, A. (1964), *The Growth of Television Ownership in the United Kingdom*, Cambridge, Cambridge University Press.
5. For example, Ray, G. (1969), 'The diffusion of new technology: a study of ten processes in nine industries', *National Institute Economic Review,* 48: 40-83.
6. Mansfield, E. (1968), *Industrial Research and Technological Innovation*, New York, Norton; Ironmonger, op. cit.

7. The accumulation of intangibles by high-technology firms
M. Teubal

Introduction

Innovation and technical change have long been recognised as significant factors in the growth of firms and industries, and significant attempts are being made to incorporate the phenomena into theory and empirical analysis. This chapter looks at empirical attempts to analyse the significance of innovation. For our purposes, it is useful to distinguish quantitative attempts at measuring contribution to productivity growth from microeconomic case-studies of individual firms or sectors. The standard procedure in the former is to posit for every sector or firm included in the analysis a relationship between explicit inputs to innovative activity—R & D expenditures—and output, profits or productivity.[1]

The above procedure ignores the variety of mechanisms of intangibles accumulation or capability creation, of which knowledge from R & D is just one possibility. It also ignores the variety of ways by which innovation affects growth. Both have been illustrated by the increasing number of case-studies of firms and industries from a number of different countries, sectors and contexts. Studies of innovative performance, such as Project SAPPHO,[2] have shown that the factors separating the successful innovations within a set of instrument innovations from the failed ones are to some extent different from those separating successful chemical innovations from failed ones. The relative importance of factors relating to marketing, understanding of user needs, and market feedbacks was found to be higher in the former set of innovations than in the latter. These differences may indicate that different industries accumulate different kinds of intangibles. Another tradition of research—one that focuses on the growth of firms through time—shows that operating experience, rather than explicit R & D, is a central factor explaining the increased efficiency of new processes in industries such as petroleum-refining and steel.[3] In metalworking firms, on the other hand, a pervasive mechanism explaining the learning and technological capacitation process is the accumulation of manufacturing abilities, with design capability and R & D appearing only at a later stage of the process.[4] It follows that it may not be appropriate to use R & D as one of the intangibles, or as the only intangible, for all branches of industry.[5]

The purpose of this chapter is to contribute towards conceptualising the various mechanisms of intangibles accumulation that have been identified in various industrial branches. It is hoped in this way to strengthen the links between case-studies and more general quantitative attempts to measure the contribution of innovation to economic growth. In addition, it may help us in understanding the indirect contribution of production, innovation and investment in a variety of branches of industry. More specifically, the purpose of this chapter is to (1) distinguish patterns observed in process industries and in metalworking from those observed in dynamic electronics areas, (2) illustrate the complex nature of intangibles accumulation within the latter, and (3) suggest some ways of representing the various patterns. Throughout the discussion a distinction will be made between the direct profitability of activities such as investment and innovation and their indirect profitability. The latter results in part from the accumulation of intangibles, including the effect of activities at any given time, t, in changing the profitability of future activities.

Intangibles accumulation in process and metalworking industries

The pattern of technical change and intangibles accumulation that has mostly been considered in case-studies relates to the process industries and to metalworking. This will be reviewed as a background to the description of the pattern observed in some firms in electronic instruments.

A. *Operating experience in process industries*

The studies by Hollander and Enos document the achievement of increased productivity and reduced unit production costs in rayon production and in petroleum-refining respectively.[6] Hollander investigated the history of Du Pont's rayon plants for periods of between 14 and 23 years, during which time unit cost declined by yearly rates of between 2.3 per cent and 4.9 per cent. An attempt was made to separate the effects of technical change from those of economies of scale (for example, the spreading of fixed costs resulting from the duplication of plants and other factors). Technical changes were classified first as major or minor (according to difficulty and the amount of investment required), and second, into those changes which reduced costs at existing levels of output (for example, waste reductions) and those where cost reductions depended on increases in output. The latter cost reductions derive either from the stretching of capacity—termed indirect technical change, as might be achieved by increasing spinning speeds of existing machinery—or from the introduction of new machinery.

Hollander's analysis showed that technical change was of overwhelming importance in explaining the observed reductions in cost, with minor changes accounting, in all rayon plants except one, for at least 75 per cent of all the reductions in cost due to this factor. Moreover, most of the changes were of the capacity-stretching type; they originated in current operations and involved the co-operation of technical assistance to production groups rather than the central

R & D laboratory. Additional microeconomic work confirmed the importance of operating experience and minor improvements in raising productivity in the context of rayon plants in Argentina and in steel plants.[7] These studies showed that this experience led to substantial capacity-stretching—minor improvements whose main objective is increased capacity rather than direct reductions in cost. They also suggest that capacity-stretching was an efficient alternative to conventional investment, although the firms involved turned to it only when forced by external circumstances.[8] In addition, a study of a big Brazilian producer of welded pipe confirms an additional fact found from the steel plant studies, namely that operating experience was a central determinant of the increased efficiency of the firm in its subsequent investment programme. The mechanisms involved include experience with the layout of the machinery, experience with the operation of existing machinery which permitted more efficient designs, and the greater capability of predicting the times and costs of erecting new plants.[9]

The above pattern of intangibles accumulation from investment and plant operation has implications for determining both the direct and the indirect profitability of activities in the process industries. Thus, the direct economic return to investments at t (cumulated operating profits net of investment costs) may have to include the effects of operating experience in reducing costs via minor improvements and capacity-stretching. In a very important sense, all of these benefits should be attributed to the original investment since the latter is the base which sustains the string of improvements.[10] In addition to the direct return, there is an indirect return to investing at t. This derives from the lower capital requirements, adjustment delay (and gestation period), and risk associated with subsequent investment programmes. They result both from the planning and execution of the original investment, and from the operating experience accumulated.

The case of a successful steel plant in Brazil. The study of the Usiminas plant by Dahlman and Fonseca provides very interesting material on intangible accumulation within this very successful steel firm.[11] The history of the firm could, in principle, be described as a succession of investment spurts or programmes where the planning, execution and operation of a programme led to the accumulation of intangibles which benefited subsequent ones. The original plant had a design capacity of 500,000 tons per annum, and was planned and installed almost wholly by Japanese minority partners in the firm. Production began in 1963, and after three years the design capacity output was achieved. Between 1966 and 1972 output increased continuously, and thanks to a succession of minor technical changes and improvements, reached a level of 1,200,000 tons per annum—a stretching of 140 per cent beyond nominal capacity. The first expansion for an additional 1,200,000 tons (the second investment spurt) was completed in 1973. The expansion benefited principally from operating experience in the original plant, and the Brazilians acquired substantial engineering capabilities from their participation in the planning and execution of this project. In fact, they performed about 40 per cent of the engineering. This experience was very important for the second expansion (the third investment spurt), which raised capacity by an additional 1.1 million tons per annum for which planning and execution was the total responsibility of the Brazilians. The company subsequently sold engineering services to other steel plants both within Brazil and abroad. The data appearing in the case-study are not sufficient to calculate the direct rate of return for each investment programme, and certainly not to calculate the value of the indirect contribution of each investment programme to subsequent ones. There is, however, substantial information on the nature and mechanisms of intangibles accumulation and it may be worthwhile to spell them out.

Operating experience. A lot of information on the original installation is available, and of how it affected the first expansion programme. The firm gradually learned how to operate the equipment and, in particular, how to select and prepare the available raw materials, a process requiring considerable experimentation, design changes in the machinery and the addition of auxiliary equipment. It also learned how to stretch the capacity of the plant, a process which also involved experimentation and design changes, although actual investments were low (the original plant with nominal capacity of half a million tons cost $261 million, while the stretching to 1.2 million tons was estimated to have cost only $40 million). Specifically, operating experience brought better specification of equipment—the new sinter machine, for example, incorporated screening and roll crushers to control sizes of the mineral ore entering the machine, while the volume of the new blast furnace was specified on the basis of the firm's experience with the selection and preparation of raw materials. Changes in auxiliary equipment and in refractory linings, some of them associated with capacity stretching in the original plant, were also specified. Similar changes were made with the specifications of the new steel shop, which incorporated both the results of past experience and exogenously available improvements. The result of all this was the erection of a more efficient plant at less cost. Both unit capital costs and unit operating costs (materials, labour and energy) of the second investment programme were lower than they would have been in the absence of the original investment. Better layout was also stated to have resulted from past experience.

Acquisition of engineering capabilities. This was particularly striking during the first expansion plan, and it gave the firm total independence from a foreign consultant during the second expansion plan. It should be noted that in the same way that the acquisition and utilisation of operating experience required explicit investments and organisational changes (such as establishing and staffing the Industrial Engineering Department and support-oriented Research Centre), the acquisition of an engineering capability required working in close collaboration with a foreign consultant and investing heavily in know-how and in training.[12] In 1970, a separate 'Superintendency' of Development was established, and by 1975 it embraced a number of separate departments (Process Engineering, Basic Engineering, Equipment Engineering, etc.). This case seems to show an interconnected sequence of intangibles accumulation, starting with operating experience in a broad sense—including knowledge of materials, process and quality control, maintenance procedures, and so on—followed by equipment design and engineering investment capabilities, and even equipment manufacturing capability.

B. Manufacturing abilities in metalworking

The original learning curve concept relates unit costs of a particular product to the accumulated output of that same product. This has been shown to be relevant in the production of a wide variety of items (metallic and others) from airframes to integrated circuits. However, case-studies of individual metalworking plants show that, in addition to the above simple learning, firms learn to produce increasingly complex or sophisticated products, for example, products with stricter specifications and/or greater weight. This phenomenon of qualitative learning is extremely important for understanding the development of infant capital goods firms in developing countries, and in particular for understanding the emergence of an export capability.[13]

There are several points concerning qualitative learning that are worth mentioning. While operating experience involving the plant as a whole dominates in the process industries, the manufacturing abilities being accumulated in metalworking firms refer to individual operations or processing steps common to a wide variety of products.[14] They involve greater skill in performing both individual operations and an increasing array of operations. It follows that the accumulation of these abilities is the basis for product diversification, and in particular for the transition from simple to complex products. (The counterpart in the process industries is the increasing ability to plan and execute complex investment programmes.) For example, the welding of overhead cranes in a heavy capital goods Argentine firm contributed to the capability of welding nuclear power components of much stricter specifications—and of obtaining orders for these products.[15]

Complementary investments required to enter the new product lines may be very substantial when the latter involve heavier or more sophisticated products. They probably involve more expensive equipment for quality control as well as new production machinery. Thus, in contrast to simple learning, intangibles accumulation is not a pure spin-off in this case. The accumulation of manufacturing abilities sets the base and increases the incentive for acquiring a design capability, first in a relatively narrow range of products. This need not mean R & D and prototype testing, but rather a capability for tailoring a particular product to specific users. (The equivalent design capability in process industries may refer to process machinery or to the capability of planning or executing investments.) Formal R & D may come later on, sometimes as a spin-off of quality control.

The case of a successful lathes producer. The work by Castano et al. is an interesting case of the kind of work which may lead to a better understanding of the main issue of this chapter.[16] The authors divide the history of a successful Argentine machine-tool producer into a number of stages, each one characterised by its organisational structure and by a set of product and process innovations. Although the study does not directly focus on the process of intangibles accumulation, it provides significant information on some aspects of this process, especially those related to the growth of design capability.

During stage I (1945-60), the firm was a family-owned artisan shop devoted to producing foreign designed lathes for the domestic market (for example, universal machines for repair and maintenance shops). There was no real design effort, merely the copying of foreign models, although these models increased in complexity over time. While the first item was copied in an artisan way (without extensive use of drawings and plans), the last item copied—the Ursus lathe (1958)—required the services of a professional design engineer.

The second stage (1960-65) began with absorption of the shop into a larger foreign-owned group which had only recently set up a small shop of its own for the supply of special lathes to the emerging car and parts producer market. That small shop's manager—a German engineer with extensive design experience—became the head of the new company, and formally set up the design department of the firm. During this period of very rapid growth, the firm copied or produced under licence a few models of special lathes for the new markets. Towards the end of the period, and in response to a decline from these segments, it launched its first product wholly of its own design—the highly successful universal T-190 lathe. This product pushed the firm to a leadership position within the Argentine machine-tool sector. It incorporated improved features, such as increased weight to sustain the higher speeds of a stronger cutting tool. Mastery of new techniques, such as heat treatment and the rectification of gears, was required. Investments in equipment, training efforts and organisational changes were critical to ensure success of the T-190 (firm output tripled during the period). Although the authors state that previous experience in copying the Ursus was important for designing the T-190, they do not focus on the role of pre-existing production skills (and possible reputation and knowledge of markets) in enabling the firm to diversify.

During the third period (1965-69), the local market slackened and the firm responded by adapting the basic T-190 model to a variety of user segments, and by widening and improving the line of special machinery. The skills and infrastructure developed for the T-190 must have been a significant contributory factor to this new effort of diversification, but more information on this is needed. The study focuses, however, on the complementary efforts undertaken to ensure that the firm maintained its previous levels of output and productivity throughout the crisis, given the wide variety of products being manufactured. These efforts included: (i) the establishment and development of separate Departments of Methods, Design, and Production Planning and Control, and (ii) the employment of more people not directly engaged in production (including technicians and engineers).

This interesting case shows that it may be possible to structure the history of the firm as a sequence, or tree, of product or product lines (Ursus, T-190, specialised lathes) linked by production, design and possibly other types of experience (spin-offs). In addition, a critical independent role should be attributed to discrete, non-project-specific investments, to changes in organisation, and to incorporation of new technical personnel. Such a firm profile may shed new light on the role of experience in productivity growth, and may be more useful, given the importance of new products, than a profile based on investment spurts, such as the one suggested for steel firms.

Intangibles accumulation in electronics

The available evidence suggests that the learning process within electronics firms differs considerably from the learning processes suggested above for the process and metalworking industries. The description that follows is based on the R & D histories of a series of electronics firms during the 1970s, and in particular on the development of a very successful Israeli instruments firm. A full report on the latter can be found in Teubal (1981):[17] only some aspects of the firm's development will be considered here. The R & D history was expressed in terms of a series of R & D projects representing successive generations of a given product, transitions to more complex substitute products, and diversification towards other products. The qualitative information suggested very strongly that the profitability of a project at time t depended considerably on the projects undertaken prior to t, and that the magnitude of this effect depended on changes in the environment, for example, exogenous availability of new technology such as new electronic components. In the mechanisms involved, production experience did not figure prominently, and R & D, although important, could only serve doubtfully as a proxy for the multitude of mechanisms involved.

The set of projects and their direct profitability

The products launched by the firm in the area covered by the case-study were grouped into nine R & D projects belonging to a total of five product classes.[18] There is more than one project for product classes II and III—four and two projects respectively—each one corresponding to a particular generation of that class. During its first year of existence, the firm was involved in three R & D projects and only one was directly profitable—the first generation of product class III. The first generation of product II was a technical failure and was never launched. The list of projects and their direct profitability is shown in Table 7.1.[19] The direct profitability measures (p_i') are based on the ratio of the present value of sales to the present value of R & D expenditures, with the base period being the year in which most of the R & D was undertaken.

Table 7.1. R & D profitability of the case firm's projects*

Project i		Base year	$p_i' = p_i/p_3$
1	I	1	0.35
2	II first generation	1	0.00
3	III first generation	2	1.00
4	III second generation	3	1.85
5	IV	4	2.89
6	II second generation	5	0.33
7	V	6	0.19
8	II third generation	6	1.06
9	II fourth generation	9	1.79†

* No significance attaches to the absolute values of these ratios. Their usefulness lies in showing the variation of project profitability within and across products.
† Underestimate.

Each point of Figure 7.1 shows the year in which most of the R & D of a particular project was undertaken (in most cases, the year of launch of the main product included in the project), and the profitability of that project. The solid line connects successive R & D projects, while the broken lines connect the various generations of products III and II. The figure shows that while the direct profitability of successive generations of product increased, the shift from one product to another may be associated with a reduction in profitability. Products IV and V represent efforts at diversification on the part of the firm, while product I groups a line of products tried by the firm at its inception and later abandoned in favour of more successful products. Products II and III involve a principal instrument and a series of optional accessories and peripherals, launched at various intervals of time and to some extent common to both product classes. Although both II and III served somewhat different users at an early stage, technological developments eventually gave II a decisive advantage over III for all users. The first generation of product II may thus be regarded as a premature attempt by the firm to innovate in this area—one that was suspended prior to attaining technical success. In general terms, successive generations of a product were more complex, commanded a higher price and involved higher R & D costs. Instruments belonging to II were also, generally speaking, more complex than those belonging to III.

The indirect profitability of projects

The pattern of rising direct profitability across generations and the qualitative evidence suggested that early generations contributed to the profitability of later generations. In other words, the former's total profitability exceeded direct profitability. In addition, qualitative evidence in projects and product classes shows that product classes launched earlier contributed to the profitability of product classes launched later on. This contribution was significant because the firm managed to adapt itself efficiently to exogenous changes in the environment, such as the availability of new components which reduced the profitability of old generations/product classes and increased the profitability of new generations/product classes.[20] What were the main factors associated with a project which benefited subsequent projects?[21]

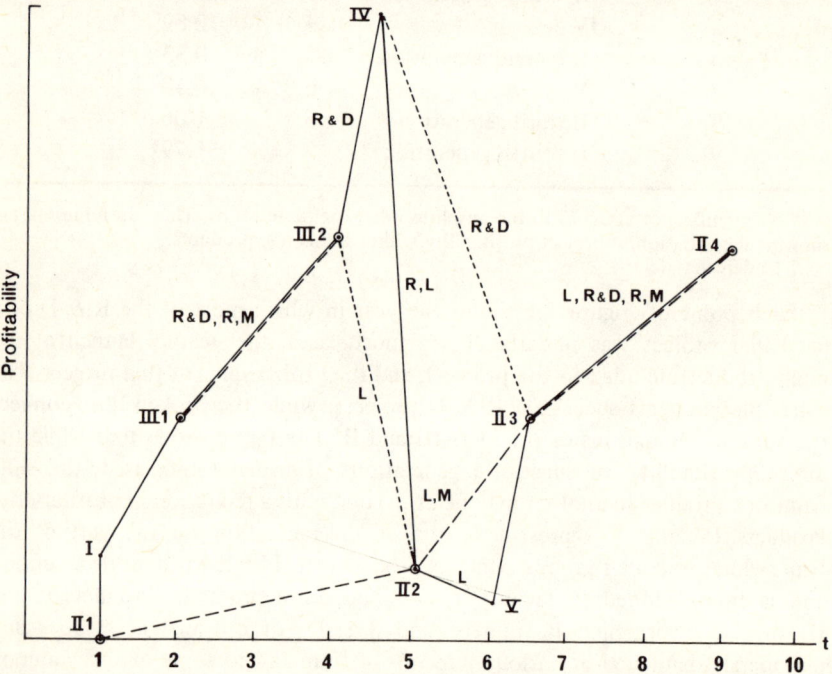

Figure 7.1. Project profitability and interrelationships (electronic instruments firm)

We may distinguish two main means by which a project benefits subsequent projects: through intangibles and through capability based on physical capital, both accumulated while executing the earlier project. The qualitative evidence is overwhelming in indicating the importance of the former group of factors and the relative unimportance of the latter. This is probably much more typical in young electronic instruments firms than in metalworking firms. Young electronics firms usually start investing in plant and equipment once a successful prototype has been tested and strong demand signals suggest that it may have a market; in particular, an order to supply a requisite amount of equipment is sometimes a requirement for investment, and this requires a pool of marketing and technical knowledge. Also, products are not standard, so cost-effectiveness based on integrated and massive production processes is not a central requirement for success.[22]

What follows will concentrate on four types of intangibles flowing from one project to another: useful technological knowledge or R & D synergy (simply termed R & D); useful knowledge obtained from marketing or from user feedbacks (M); a line effect (L), which may be one-way when the earlier project benefits the later project, or two-way when, in addition, existence of the later project increases sales of the earlier project; and firm reputation effects (R) which generally, but not exclusively, result from past sales. In the firm under discussion, R & D includes instances where a whole sub-assembly developed for a particular generation is the base upon which improvements on it are incorporated into subsequent generations. M includes instances where user feedback or marketing efforts have shown the desirability of changing product specifications, or the desirability or possibility of offering additional products. L will exist whenever old customers of the firm are more likely than new customers to purchase an 'add-up' to the instrument purchased in the past, or another instrument which complements the one purchased in the past in some way (due to familiarity in use and in maintenance, ease in interfacing, etc.), or when new customers or dealers have a strong preference for purchasing from a firm offering a complete line of products rather than from firms offering individual products only.[23] Finally, R is generally acquired from past sales of similar or related products which have satisfied customers from a functional point of view, but it also includes an instance where the firm acquired a reputation for technical excellence and sales potential in the eyes of a prospective sales agent.

The pattern of relationships observed

A 'minimum' pattern of the more pronounced project interrelationships is shown in Figure 7.1, where the symbols for the various effects are located over the segments connecting two projects. Most of the symbols are located on segments joining two consecutive projects. The important exceptions are shown by the dotted lines connected III2 to II2 (the second generation of products III and II) and IV to II3 (product IV and the third generation of product II). The main points to note are:

1. There are a number of types of intangibles accumulation, only one of which is R & D. The accumulation of production experience across projects did not

seem to be very important, although the experience curve operated within individual products.
2. Relevant intangibles benefited both subsequent generations of a product class, and also the design and sale of new product classes. Thus product obsolescence was much stronger than intangibles obsolescence. Even when direct benefits exist potentially, their actual realisation depends on the capacity of the firm to shift successfully to new products, and thereby assure its own growth and profitability. This requires, in addition to the intangibles accumulated, significant investments, principally in R & D and in marketing.
3. The indirect benefits of the project not leading to sales (the first generation of product II) were low or non-existent, even with respect to R & D. In general, it seems that little information on the targets of R & D for subsequent generations can be accumulated without actual sales and user feedback.

Issues and problems

Intangibles from inputs or outputs

We have seen that a number of mechanisms for intangibles accumulation in industry operate, and it may be useful to summarise them by distinguishing those which flow from the inputs to particular activities or projects from those which flow from R & D (input to particular innovation projects), while knowledge about markets is generally associated with another input—marketing. This picture is evidently too simplistic because it abstracts critical intangibles obtained from users—market feedbacks—an important part of which depends crucially on sales, that is, on outputs of innovation projects.[24] Similarly, the experience and knowledge relevant to investments depend on both inputs (participation of a firm's personnel in the planning and execution of investments) and on outputs (operating experience of the new plant). There are, however, some intangibles which may be associated with either input or output. For example, production skills and firm reputation, which generally depend on inputs and on outputs (successful sales of new products) respectively. Finally, several studies make reference to the accumulation of design skills in firms, particularly in the metalworking and machinery sectors, rather than the accumulation of technical knowledge. What factors explain this process? It may be tempting to distinguish between pure design skills—a capability to design products of increased technical performance relative to existing products—and a capability to produce designs which improve the utility or service provided to users (relative to cost) and which can be manufactured by the firm concerned. The former does not necessarily lead to the latter, as is already well known from innovation studies.[25] Although pure design skills may arise from past experience with purely technical activities and from current R & D, effective design skills—those which in principle can lead to commercially successful designs—depend also on market knowledge and on knowledge about manufacturing capabilities.[26]

Spin-offs versus explicit resource allocation

It is customary to distinguish intangibles which are by-products from another activity from intangibles generated by a purposeful allocation of resources. The typical example of the former is production experience: that of the latter is technical knowledge acquired through R & D. It may be that this distinction is less useful for understanding the development of high-technology firms than one based on identifying sequences of activities or projects, and the links (spin-offs) between them. Thus, the contribution of 'doing' activity A to activity B may include intangibles accumulated both from doing and from explicit resource allocation, provided that they were specific to project A. Elsewhere the relevant earlier activity or projects have been identified as product innovation projects, and the contribution was termed a spin-off.[27] A closer look at that case, however, and at other firms, shows the appropriateness of considering other explicit investments in R & D and in quality control procedures (or explicit organisational changes) which, while not specific to a particular innovation, will have an effect on the profitability of later innovations. For example, establishment costs of an R & D laboratory or of an export marketing infrastructure would fall in this category. In one electronics firm, the costs of setting up the technical and other infrastructure for a new product class (II) explains the reduction in direct profitability observed in the short-run when shifting to that class.[28] This cost includes the need for mastering new technologies required for the newer and more complex product class, and physical investments. Therefore, the cost should not be attributed to the first generation alone.[29] Under the circumstances the profits of subsequent generations involve both a spin-off from past generations and a component attributable to past investments not specific to a product class.[30]

The economic value of intangibles

Very few attempts at measuring the value of specific intangibles accumulated by firms have been made, because of conceptual and data problems. In some studies, such as those of Hollander and Enos, a great deal of information has been collected and this could be used, in principle, to calculate the rates of return. It seems that part of future effort at the micro-level should be directed towards this objective. The purpose here is to report on some preliminary calculations made for the electronics firm referred to earlier, and to provide some relevant, though scant, information from two Brazilian capital goods firms. Some of the factors which tend to make the value of intangibles high or low will then be indicated.

Table 7.2 reproduces the calculations performed for the electronics instruments firm. The indirect profitability of the first generation of product III was assumed—on the basis of qualitative evidence—in its entirety to have benefited the second generation of III and an essentially similar procedure was carried out

Table 7.2. Direct and indirect profitability (electronic instruments firm)

	Assumptions*	
	(1)	(2)
Profitability of product III first generation		
1. Direct	0.16	0.25
2. Total: direct *plus* second generation profitability attributable to first generation	0.30	0.63
Profitability of product II related projects†		
3. Direct	0.17	0.24
4. Total: direct *plus* fourth generation profitability attributable to prior projects†	0.24	0.46
Indirect profitability as per cent of total		
Product III (line 2 − line 1)/line 2	47	60
Product II (line 4 − line 3)/line 4	29	48

* Let m be average mark-up of unit variable costs and α the ratio of R & D to total fixed costs for the project. The value of m is arbitrarily set in column (1) at $\frac{1}{2}$ and in column (2) at 1. The ratio α is assumed to be 0.5 in both columns.

† Comprises the second and third product II projects and project IV. The base year was arbitrarily set at the base year of the latest of the projects in the set, the third product II project.

for product II. It was also assumed that the later projects would have been undertaken in the absence of the earlier projects, and that their discounted profits would have been zero (which in this case is an overestimate). The indirect profitability computed as a share of direct plus indirect profitability varied between 47 per cent and 60 per cent for product III, and between 29 per cent and 48 per cent for product II, depending on the assumptions made concerning mark-ups and the share of R & D in total fixed project costs. These calculations are only of illustrative value, but they are consistent with the view of managers that inter-project spin-offs were extremely important in the development of the firm. The numbers obtained are underestimates, although the contribution of early innovations to later ones has not been separated from the contribution of early non-specific investments in R & D and infrastructure.

The data collected from the two Brazilian firms interviewed and the data reported in other studies do not permit ready calculation of the value of accumulated intangibles. A major Brazilian producer of sugar-processing and alcohol-production equipment calculates that its past activity in the field is expected to save roughly $22 million in royalty payments, but that there are also other benefits derived from production experience. Another Brazilian firm, a successful pipe producer, stated that the value of the firm's experience was very large, but that it cannot be estimated from the company's accounts. It derives from shorter lead times in establishing new plants and in training the labour force, and from saving in royalty and other payments for technological transfer.[31] Finally, the value of some of the investment and operating experience of the Usiminas steel plant of Brazil and of an Acindar steel plant in Argentina is probably very high since the significant capacity-stretching undertaking would have a direct impact on the profitability of subsequent investments.

A number of factors explain why the value of spin-offs can be high. In our electronic instruments firm, exogenous changes in the environment—especially the availability of new technology—coupled with a capacity to adapt to these changes have produced the effect mentioned. It should be noted that this case was characterised by a high rate of product obsolescence and also by a high rate of knowledge complementarity between old and new product classes. Thus, if a firm is willing to take the necessary risks by investing substantial resources in R & D and in marketing, it seems to stand a chance of maintaining or increasing its performance. Unwillingness to commit substantial resources to new projects which exploit newly available technology is probably a major reason for a low realised value of intangibles, especially in areas where prior knowledge, experience and capability may be extremely important. In other cases, the science, technology and experience underlying the new products appearing in the market may be completely different from those underlying current products. In these circumstances, the potential value of the accumulated intangibles is low and the firm may opt to get out of the market. Finally, the value of intangibles may be affected by unexpected reductions in the size of the market and by government policies. For example, a design team embodying the firm's pool of design experience may have to be dissolved due to unexpected reductions in orders. These may result from macroeconomic policies such as reduced government procurement or currency overvaluation.

Are patterns of intangibles accumulation industry-specific?

The patterns of intangibles accumulation presented above suggest that they are industry- or branch-specific, that is, a pattern based on investment experience was associated with the process and steel industries, one based on production experience was associated with metalworking, and one based on technical and market knowledge was linked with instruments and electronics. This association should not be overemphasised; it is probably more relevant for young firms than for big corporations and, to some extent, to firms in developing rather than developed countries.[32]

Thus, production experience is very important for electronic components and may be very important for producing certain kinds of electronic products, such as flat TV sets, where assembly is not the only or the main step. Similarly, the predominant accumulation of manufacturing skills in metalworking firms is probably more typical of firms active in protected markets, and of young firms in developing countries. It is conceivable, however, that even in metalworking one can find young firms involved first in design and then in production (a pattern commonly found in electronics), especially if the original founders acquired production experience from employment in other firms.[33] Finally, in metalworking industries, intangibles associated with R & D and design may be of extreme importance, even in sectors which have traditionally been considered to be conservative, such as textile machinery. Rothwell has shown that non-innovative firms in this area tended to fail despite their excellent experience and capability in manufacturing.[34] The increased application of new chemical, electronic and aerodynamic technologies resulted in a series of radical innovations which revolutionised the products of the industry. Thus, production experience, while still important and even a necessary condition for success, was no longer sufficient. New types of technological and market knowledge were becoming relevant, some of which required explicit R & D and the establishment of an R & D capability. This may also be the case in other metalworking and machinery branches.

Implications for innovation performance

The learning perspective to the development of high-technology firms, with its focus on the links and spin-offs between early activities or projects and later activities or projects, may also be useful for analysing innovation performance at the project level. The approach usually followed in the innovation performance tradition is to identify factors differentiating successful from failed innovations. For example, failed innovations may have resulted from insufficient investment in R & D, lack of interaction with customers, or management failure to take into account other critical factors for success. The projects and most of the variables are usually not dated. The learning perspective would state that, in some cases, management may have attempted to act in the appropriate direction and may even have invested considerable resources. However, due to insufficient knowledge and experience (intangibles) accumulated from the past, the efforts of the firm were inefficient. The focus is on project selection, on whether the firm's response to the new requirements or opportunities fits in with its past experience (and capabilities). On the other hand, the commercial success of a particular innovation could be as much the result of the specific tactics adopted as of thoughtful project selection which made full use of the intangibles and capabilities accumulated from the past. Thus, the analysis of any particular innovation is embedded in a view of the firm's development through time.

Conclusions

With some notable exceptions, case-studies of firms do not directly analyse the accumulation of intangibles, although they do refer to the role of experience in very general terms. Only one attempt has been made to measure the value of experience. Firm studies focusing on the accumulation of intangibles should: (i) identify the main activities (investment spurts, products or R & D projects), tracing the development of the firm, (ii) define and compute the direct profitability of such activities, and (iii) specify the main types or mechanisms of the accumulation of intangibles and attempts calculating their economic value. A main aspect of (iii) refers to the sequence of capabilities —from copying to wholly in-house designing, from production experience to the accumulation of a design capability. In particular, how does production/operating experience increase a design capability? A number of mechanisms have been introduced: preventive maintenance activities, which require intimate knowledge of the machinery (supply and demand factors); experience with available raw materials, which helps to specify targets for machinery modifications and adaptations; and similarly, use of specific machinery items, which indicates the desirability of designing (or copying) and producing such machines in the future. Moreover, quality control procedures enhance knowledge of materials and testing procedures, and therefore pave the way for more basic R & D connected with materials.

A learning perspective to the development of high-technology firms may also contribute to identifying the relevant variables in empirical productivity, profit or export equations. Cumulative R & D cannot by itself represent the process of intangibles accumulation, even in electronics, where, to some extent, it is the 'engine of growth'. This, apparently, is particularly true for the R & D associated with failed innovations. Both in electronics and in other industries, performance is highly dependent on intangibles other than those flowing from R & D, such as market knowledge, experience from investing, firm reputation effects, and so on. The selection of an appropriate variable or variables depends on the industry or set of industrial branches considered, and on the number of observations.

Production experience is probably not well represented by cumulative past output since an important part of it is increased capability to produce more sophisticated products. The main variables which would change skills and capabilities in this area are the level of sophistication achieved and explicit investments in purchasing and absorbing new technology. In dynamic product areas, such as electronics, it is extremely important to consider explicitly the market creation and the competitive pressures effect of newly available technology. The capacity of a firm to adapt to these changes depends on the pool of intangibles available from the past, and on current investments in R & D marketing infrastructure. These factors will determine the extent to which the product profile of the firm will match the optimum profile, where the latter depends on technology and on competition. The next task is to suggest formal representations for these intangibles.

Notes and references

1. See Griliches, Z. (1980), 'Returns to research and development expenditures in the private sector' in Kendrick, J. and B. Vaccara (eds), *New Developments in Productivity Measurement and Analysis*, National Bureau of Economic Research, Chicago, University of Chicago Press.
2. See Science Policy Research Unit (1972), *Success and Failure in Industrial Innovation: Report on Project SAPPHO*, London, Centre for the Study of Industrial Innovation; Rothwell, R. et al. (1974), 'SAPPHO updated. Project SAPPHO phase II', *Research Policy*, 2: 258-91.
3. See Enos, J. (1962), 'Invention and innovation in the petroleum industry' in Nelson, R. (ed.), *The Rate and Direction of Inventive Activity: Economic and Social Factors*, National Bureau of Economic Research, Princeton, N.J., University of Princeton Press; Hollander, S. (1965), *The Sources of Increased Efficiency: A Study of the Du Pont Rayon Plants*, Cambridge, Mass., M.I.T. Press; Maxwell, P. (1977), *Learning and Technical Change in the Steel Plant of Acindar, S.A. in Rosario, Argentina*, IDB/ECLA Research Program of Studies in Science and Technology, Working Paper No. 4, Buenos Aires; Katz, J. et al. (1978), *Productivity, Technology and Domestic Efforts in Research and Development—the Growth Path of a Rayon Plant*, IDB/ECLA Research Program of Studies in Science and Technology, Working Paper No. 13, Buenos Aires; Dahlman, C. and F. Fonseca (1978), *From Technological Dependence to Technological Development: the Case of the Usiminas Steel Plant in Brazil*, IDB/ECLA Research Program of Studies in Science and Technology, Working Paper No. 21, Buenos Aires. R & D may, however, still be critical for the launching of new processes.
4. See Teubal, M. (1982), *The Role of Technological Learning in the Export of Manufactured Goods: the Case of Selected Capital Goods of Brazil and Argentina*, Interamerican Development Bank; Castano, A., J. Katz, and F. Navajas (1981), *Etapas Historicas y Conductas Technologicas en una Planta Argentina de Maquinas Herramienta*, Programa BID/CEPAL/PNUD de Investigaciones sobre Desarollo Cientifico y Technologico en America Latina, Monografia de Trabajo No. 38, Buenos Aires. Firm reputation effects were also found to be very important within the Brazilian and Argentine capital goods firms covered in Teubal, op. cit.
5. Another criticism of the production function approach is that it ignores the relationship: (i) between intangibles accumulation and the accumulation of physical capital, and (ii) between economic growth and innovation. The latter has been emphasised since Schmookler, J. (1966), *Invention and Economic Growth*, Cambridge, Mass., Harvard University Press.
6. Hollander, S. (1965), *The Sources of Increased Efficiency: A Study of the Du Pont Rayon Plants*, Cambridge, Mass., M.I.T. Press. Enos, J. (1962), 'Invention and innovation in the petroleum industry' in Nelson, R. (ed.), *The Rate and Direction of Inventive Activity: Economic and Social Factors*, National Bureau of Economic Research, Princeton, N.J., University of Princeton Press.
7. See Katz, J. et al. (1978). *Productivity, Technology and Domestic Efforts in Research and Development—the Growth Path of a Rayon Plant*, IDB/ECLA Research Program of Studies in Science and Technology, Working Paper No. 13, Buenos Aires; Dahlman, C. and F. Fonseca (1978), *From Technological Dependence to Technological Development: the Case of the Usiminas Steel Plant in Brazil*, IDB/ECLA Research Program of Studies in Science and Technology, Working Paper No. 21, Buenos Aires; and Maxwell, P. (1977), *Learning and Technical Change in the Steel Plant of Acindar, S.A. in Rosario, Argentina*, IDB/ECLA Research Program of Studies in Science and Technology, Working Paper No. 4, Buenos Aires.
8. For example, by an unexpected financial constraint blocking the way to conventional expansion.
9. See Teubal, M. (1982), *The Role of Technological Learning in the Export of Manufactured Goods: the Case of Selected Capital Goods of Brazil and Argentina*, Interamerican Development Bank. Capacity-stretching of existing plant would be one

outcome of this experience, one that is: (i) disembodied, and (ii) can be incorporated into the existing plant. The remaining experience would benefit only new plants.

10. In other words, the direct return to investment at t should include a portion of the indirect return or spin-off from current production. If operating experience does not translate automatically to costs—if other investments in skill, etc. are required—then only a part of the operating profits resulting from lower costs should be attributed to the other (complementary) investments.
11. See Dahlman, C. and Fonseca, F. (1978), *From Technological Dependence to Technological Development: the Case of the Usiminas Steel Plant in Brazil*, IDB/ECLA Research Program of Studies in Science and Technology, Working Paper No. 21, Buenos Aires.
12. Maxwell mentions how another aspect of operations—preventive maintenance—enhanced the design capabilities of the Argentine steel firm which he studied. The overall picture of intangibles accumulation of this firm resembled the one referred to in the text.
13. See Teubal, M. (1982), *The Role of Technological Learning in the Export of Manufactured Goods: the Case of Selected Capital Goods of Brazil and Argentina*, Interamerican Development Bank.
14. See Rosenberg, N. (1976), *Perspectives on Technology*, Cambridge, Cambridge University Press, Ch. 1.
15. Product diversification also takes place and may even be easier within a particular product class. This particular firm launched a series of increasingly heavy overhead cranes, starting with one for in-house use, and including units for ports, the steel industry, and for hydroelectric dams.
16. Castano, A., J. Katz, and F. Navajas (1981), *Etapas Historicas y Conductas Technologicas en una Planta Argentina de Maquinas Herramienta*, Programa BID/CEPAL/ PNUD de Investigaciones sobre Desarollo Cientifico y Technologico en America Latina, Monografia de Trabajo No. 38, Buenos Aires.
17. Teubal, M. (1981), *The R & D Performance Through Time of Young, High-technology Firms: Methodology and an Illustration*, Maurice Falk Institute for Economic Research in Israel, Discussion Paper No. 814. Forthcoming in *Research Policy*.
18. These were the main areas of the firm during its first ten years of existence.
19. Further details on the groupings of products into projects, and on measuring the direct profitability of the latter can be found in Teubal, M. (1981), *The R & D Performance Through Young, High-technology Firms: Methodology and an Illustration*, Maurice Falk Insitute for Economic Research in Israel, Discussion Paper No. 814. Forthcoming in *Research Policy*.
20. Changes in the environment were thus not unambiguously favourable or unfavourable to the firm.
21. In *The R & D Performance Through Time of Young, High-technology Firms: Methodology and an Illustration* (Maurice Falk Institute for Economic Research in Israel, Discussion Paper No. 814), I mentioned experience, infrastructure and firm reputation. My purpose here is to be more specific, both with respect to the various components of the pool of intangibles and with respect to the individual projects which benefit from them. The main source of information was the 220-page file of the firm, which contains a description of each project. The information is not necessarily complete, since the interviews were not closed ones which systematically covered the various projects; nevertheless, I do believe that they indicate the main influences at work.
22. Production is largely based on relatively easy assembly operations, and involves short runs.
23. A two-way L-effect also occurs within projects and not only across projects. This is because optional accessories and peripherals are grouped together with the main instruments with which they are associated.
24. The evidence on this is very wide, especially in the machinery, metalworking and electronics sectors and, more generally, in sectors with substantial product heterogeneity, such as pharmaceuticals. It is generally considered to be stronger with respect to market knowledge, but it may be important with respect to technological knowledge

as well. Technologically sophisticated and progressive users in some industries are capable not only of providing information on product performance and abstract user needs, but also of helping to translate these needs into technical specifications, and even solving some technical problems.
25. One characteristic of the commercial failures reported in Project SAPPHO which differentiates them from innovations which succeeded was a relative lack of understanding of user needs. In general, this may lead to the absence of important product features or to the existence of unwanted features which may reduce overall performance (for example, excessive sophistication or even novelty). Some innovations covered by Teubal, M., N. Arnon, and M. Trachtenberg (1976), 'Performance in innovation in the Israeli electronics industry: a case study of biomedical electronics instrumentation', *Research Policy*, 5: 354–79 failed for these reasons. Rothwell's work also reports cases of failure due to lack of 'manufacturability' of newly designed products. See Rothwell, R. (1976a), 'Picanol Weefautomation: a case study of a successful textile machinery builder', *Textile Institute and Industry*, March; Rothwell, R. (1976b), *Innovation in Textile Machinery: Some Significant Factors in Success and Failure*, Occasional Paper Series No. 2, Science Policy Research Unit, University of Sussex.
26. There is also some evidence that manufacturing skills also make a direct contribution to design capabilities; for example, via knowledge of materials, testing procedures and techniques associated with quality control. The introduction of preventive maintenance procedures to ensure better operation of an Argentine steel plant, for instance, enhanced the firm's capabilities to modify and adapt machinery.
27. Teubal, M. (1981), *The R & D Performance Through Time of Young, High-technology Firms: Methodology and an Illustration*, Maurice Falk Institute for Economic Research in Israel, Discussion Paper No. 814. Forthcoming in *Research Policy*.
28. The actual shift is between the second generation of product class III and that of class II, the latter being the first II innovation which reached the market.
29. Considerations of this type have always been made with respect to investments in physical capital.
30. Strictly speaking, the indirect contribution of the earlier project to the later one—after having separated those expenditures which are clearly common to both—should not in its entirety be considered a spin-off. For example, operating profits of the earlier generation may have been low because the firm reduced prices to push sales in order to acquire reputation and feedback which would bring it large profits in subsequent generations. Part of the latter's higher profits were thus planned and not obtained from nothing, so it should not be termed spin-off. Whether the indirect profitability or contribution of an activity is or is not a spin-off depends not only on the nature of the activity, but also on whether the firm was looking only at short-run profits or at both short-run and long-run profits.
31. The firm also stated that its experience gave it enhanced planning capabilities which a newcomer would not possess.
32. This reflects the fact that most cases of firm learning processes seem to have been undertaken in developing countries.
33. The story of the Belgian firm Picanol seems to conform to this pattern. See Rothwell, R. (1976a), 'Picanol Weefautomation: a case study of a successful textile machinery builder', *Textile Institute and Industry*, March. In general, firms in developed countries may benefit from past experience of the industry (externalities) and therefore the learning in manufacturing need not be critical for new metalworking firms. It is, however, the critical intangible being accumulated at the eary stage of development of the metalworking sector.
34. Rothwell, R. (1976b), *Innovation in Textile Machinery: Some Significant Factors in Success and Failure*, Occasional Paper Series No.2, Science Policy Research Unit, University of Sussex.

8. Information economics and technological change

D. McL. Lamberton

Can the new information economics contribute to the understanding of the complex process of technological change? This chapter will discuss the emergence of information economics and attempt to show that its perspectives have not only contributed to such an understanding, but may also be proving instrumental in the creation of an economics of change that addresses directly the trouble with technology experienced by the modern world.

The emergence of information economics[1]

The emergence of information economics can be seen as a response to the deficiencies of economic theory built upon perfect knowledge, the failures of policy, or the spectacular advent of intelligent electronics with the greatly enhanced capacity for communication, computation and control. Whichever is the preferred interpretation, it remains a matter for personal judgement whether the battle for recognition and respectability has only just been joined, is well advanced, has been won, or perhaps has been lost.

There are numerous milestones. In 1976 the words 'and information' were added to the entry 'Economics of Uncertainty' in the *American Economic Association Index of Economic Journals*. Pioneering contributions, both theoretical and empirical, have been made over the last half century and more by, for example, Knight, Hayek, Jacob Marschak, Daniel Bell, Machlup, Shackle, Boulding, Kornai, Simon, and Arrow.[2] More recently there have been surveys[3] and symposia and conferences.[4] There is a burgeoning literature that reaches to every category of the AEA classifications.[5] As their titles suggest, some defy such processing,[6] while many address specific topics. All represent a challenge to orthodoxy. The tedious bibliographic detail serves to emphasise the pervasiveness of that challenge, to which might be added another: information economics is both bringing a new dimension to old interdisciplinary links, as between economics and law and creating new interdisciplinary links, as between economics and information science. To the extent that this spurs new territorial ambitions on the part of economists, they should perhaps reflect on their motivation. Is it 'that economists are looking for fields in which they can have some success',[7] or is there a growing realisation that the role of information may be the key to real advancement of understanding economic phenomena?

Informational efficiency has been identified as a policy objective and information resource management is seen as a method of achieving other policy objectives. Much of the modern thrust for this development came from the science sector, which all too often failed to appreciate that STI was only one of many information inputs.[8] The wisdom of the Piganiol OECD Ad Hoc Group on Scientific and Technical Information has not always been heeded. Their report took the viewpoint of information specialists, but added that 'the Group has consistently tried to identify the way in which the "information system" articulates with the structures of economic growth and, more broadly still, with the nature of political decisions'.[9] The policy debate has spread far beyond the OECD. In UN agencies, and UNESCO in particular, the New International Information Order has been debated hotly. The policy needs for transition to the information society continue to be explored.[10]

As to the scope of national information policy, Dunn advocates boundaries that include 'what we know as a people, what we are doing to learn more, and the tools that we use to conduct individual transactions and to communicate on a person-to-person basis'. He argues that:

> There is a value in bringing together the ideas and issues involved in this set of national activities, because doing so calls to our attention the interrelatedness and importance in our lives. There are opportunities for improving the operation of the systems that provide information services ... Many of these opportunities will be enhanced by taking an integrated view of this area.[11]

The Japanese Fifth Generation Computer Project seeks to implement such an integrated view. Information processing systems are expected to play the following roles: '(1) To increase productivity in low-productivity areas; (2) to meet international competition and contribute toward international cooperation; (3) to assist saving energy and resources; and (4) to cope with an aged society.'[12]

If a copious literature and admission to the policy arena are not sufficient evidence of the emergence of information economics, it might be contended that it has a history.[13] Robbins is one of few to consider the longer history of the treatment of the advancement of knowledge in economic thought.[14] He concedes 'the neglect of knowledge',[15] 'as regards emphasis in the formal structure', but feels 'that the picture is somewhat one-sided. It is just not true that the economic thought of the past was unaware of the relevance to development of technical or other forms of knowledge or that there is a lacking in the literature conspicuous emphasis on its importance.'[16] He looks to the writings of Bentham, Babbage, Charles Knight, Rae, William Ellis, Harriet Martineau, the Mills, McCulloch, Senior and Marshall. He might well have added Dugald Stewart[17] and Thomas Hodgskin; Chapter II of the latter's *Popular Political Economy* (1827) begins: 'Influence of knowledge not noticed by economists till very lately ...'.[18]

Recognition of the importance of knowledge is only a first step toward an information economics. Early writers do not appear to have gone very far in the attempt to treat changes in knowledge as endogenous, although Williams' suggestion that this began with Machlup must be questioned.[19] Knight gave careful consideration to methods of meeting uncertainty. He observed that 'information was one of the principal commodities';[20] that 'vast sums of public money', 'great investments of capital and elaborate organisations' were devoted to information activities. Admittedly he thought that they called for discussion, in the context of *Risk, Uncertainty and Profit*, 'only in so far as they affect the general outline of the social economic structure'.[21] He pointed to 'the existence of highly specialized industry structures performing the functions of furnishing knowledge and guidance'.[22] (Perhaps his comments did not attract as much attention as those of Machlup because he made no specific mention of the academic role!) Neither early writers nor Knight gave attention to the communication processes that were involved. Even into modern times, communication concepts have remained primitive,[23] two fundamentally different concepts being distinguished: *vertical* between individual actors and a central agent, and *horizontal* between actors.

Before pursuing Knight's question about the way in which methods of dealing with uncertainty affect 'the general outline of the social economic structure', some consideration of definition is called for. Spence contended that recent interest in, for example, computers and telecommunications had obscured both 'the pervasive influence of information in many other markets' and 'the impact of the information sectors on the rest of the economy'.[24] This was being redressed by recent research which focused 'largely upon the informational stucture of markets'.[25]

Stiglitz, in his introductory remarks to the Stanford symposium, again emphasised the market context, but took the view that imperfect information altered 'the conventional notion of a market'.[26] Central statements from competitive theory became questionable. 'The basic character of how we ought to view the competitive economy is altered if we take seriously imperfections of information', he said.[27] All papers in the symposium related to 'the problems arising from costly information'.[28]

Hirshleifer and Riley drew a distinction between the economics of uncertainty where individuals are limited 'to terminal actions, permitting them only to *adapt* to uncertainty' and the economics of information which examines 'the consequences of informational actions, which allow [individuals] to *overcome* uncertainty'.[29] Their survey captured a wide range of information activities: acquisition, dissemination, R & D, espionage and monitoring, wagering and speculation, signalling, and education, ending with comments on rational expectations and informational efficiency. Institutional aspects barely rated a mention. There are 'interesting complications . . . when the informational decision process has multipersonal aspects',[30] but these were not pursued.

Justification for limiting property rights in ideas were admitted.[31] It was concluded that: 'Information generation is in large part a disequilibrium-creating process, and information dissemination a disequilibrium-repairing process.'[32] Where does organisational change of the kind plainly envisaged by Knight allow individuals to *overcome* uncertainty?

Arrow adds this new dimension. The last three decades have been preoccupied with decision-making under a given uncertainty. 'The problems of the economics of information proper arise when the probability distribution of states of the world is a variable', he says.[33] This opens up ways of improving decision-making. First, the decision-maker can take advantage of the existence of signals. Secondly, the choice of which signals to receive becomes a decision variable. Thirdly, in sharp contrast to the individual decision-maker who dominates the economics of information literature reviewed by Hirshleifer and Riley, a group or organisation will normally be the real world decision-maker. This is because, as Arrow says: 'Specialization in information-gathering is one instance, in [his] view the most important instance, of the economic benefits of organization.'[34] To the extent that organisational structure and efficiency are not constrained fully by past events, organisational change becomes a third strategy for overcoming uncertainty and at the same time improves the prospects of the two other ways. Larger resources make possible fuller exploitation of the existence of signals, while organisational capacity can widen the range of signals considered for reception. However, organisational structure may be inappropriate and 'sclerosis of organizations may be as dangerous to health as sclerosis of the arteries'.[35] Arrow argues:

> that the combination of uncertainty, indivisibility, and capital intensity associated with information channels and their use imply (a) that the actual structure and behavior of an organization may depend heavily upon random events, in other words on history, and (b) the very pursuit of efficiency may lead to rigidity and unresponsiveness to further change.[36]

There would appear to be major implications for economic theory. The Hayekian approach, which emphasised that the economy was an information system but held that there were important limitations on the availability of information, left the institutional structure intact, with the consequence—for the most part not mentioned—of probable advantage to those individuals and organisations 'having access to better information, or a better position in the institutional structure'.[37] This is equally true of the original version, the modernisation by Sowell, or the resurrection by the rational (consistent) expectations approach as reviewed, for example, by Begg.[38] This latter approach prompted the comment: 'It would be ironic if the outcome of a decade and a half of concern with the economics of (imperfect) knowledge resulted in the renewed dominance of models incorporating perfect knowledge assumptions'.[39] The biggest loss might well prove to be the inhibiting of the attempt by economists to probe deeply into the organisational aspects of the economy.

Of course, it might be argued that economics, in order to cope with disequilibria, is devoting more attention to how decisions are made. Various approaches to procedural rationality have been developed: operations research and management science, artificial intelligence, computational complexity, and cognitive simulation.[40] These approaches appear to be concerned with 'the distinction between the perceived environment and the environment',[41] or, as Simon put it: 'Procedural rationality is the rationality of a person for whom computation is the scarce resource.'[42] He argues that there is no theorem that proves the process will converge to limit survival to those who have found the objective optimum:

> It is much more likely, in a world with rapidly advancing human knowledge and technology, with an unpredictable shifting political situation, with recurrent and unforseen (if not always unforseeable) impacts of demographic, environmental, and other changes, that the location of the objective optimum has little relevance for the decision-makers or the situations that their decisions create.[43]

Simon believes normative microeconomics, the theory of the business cycle, and the Schumpeterian domain of long-term dynamics are specific domains that can benefit from the procedural rationality approach. This may be beyond challenge, but some queries do arise. Consider the long term. Here the contention is that: 'The search for new products or new marketing strategies surely resembles the search for a good chess move more than it resembles the search for a hill-top.'[44] Even if this is correct, it remains possible that the decision strategy bears little resemblance to the search for a good chess move once it is recognised that the strategy can take advantage of the existence of signals, that the choice of signals to receive has to be made, and that the organisation itself is a variable.

The study of organisational change and organisational options should then be properly an activity for economists. Comments and efforts here are of various kinds. For example, in the context of industrial economics Bradburd and Over treat an industry's monopoly price as a 'collective good' achieved and sustained by organising.

> The costs of the necessary organization can be divided into those incurred in the formation of the organization and those incurred in maintaining it. In order to establish the communication demands, signalling mechanisms, and other rules of behavior which govern the operation of an organized industry, and in order to determine initially how the costs and benefits of the collective good are to be apportioned, an industry must incur fixed formation costs ... Maintaining the organization to ensure adherence to the price and output policies agreed to when the organization was formed is also costly. Resources must be expended to maintain channels of communication and on other interfirm activities such as exchanging information and policing the industry agreement, as well as on all intrafirm communication that guides the

marketing department in its efforts to compete vigorously without violating the industry agreement.[45]

The existence of organisational capital leads to separate integrative and disintegrative levels of concentration. Teubal likewise assigns an important role to 'intangibles accumulation or capability creation, of which knowledge from R & D is just one possibility'.[46] Both these efforts are in the tradition of the pioneering work of P. W. S. Andrews, who distinguished between technical and managerial costs and examined the 'Effects of Changing Organization'.[47]

At the international level, supranational forms of organisation have received too little attention from economists: one must turn to journals of international business and even anthropology.[48] While transborder data flow studies have tended to emphasise an underlying free trade vs. dependence conflict, they have also begun to cast new light on specialisation in and location of the activities of multinationals. Two intriguing aspects of current developments in communications technology might be noted. These developments challenge the new orthodoxy according to which 'technology ... has never been so internationally mobile. It now seems that only labor is a geographically fixed factor.'[49] In some ways these international firms might be viewed as organisations for international co-operation which then brings them within the scope of efforts to compare the efficiency of alternative forms.[50]

These few samples of recent research initiatives illustrate the potential importance of organisational change at the level of the firm,[51] and the industry both nationally and globally. The terminology differs: organisational capital, information stock, capability, intangibles.[52] In each case a form of investment is involved: investment that allows for the influence of past events on present decisions.

How does organisational change come about? What is the role of information in that process? One response can be to proceed as has commonly been done with technology: to assume that there is an organisational manual and that an optimal choice of organisational form will be made from the large number of alternatives listed. However, this makes no more sense for choice of organisation than for choice of technology.[53] The pages of such a manual are widely scattered, they are costly to locate and translate, and they have very limited application to the special circumstances of the existing organisation, constrained as it is by its organisational assets.

Schotter's attempt to provide an economic theory of social institutions may in due course help in answering this question.[54] Institutions are regarded as regularities in the behaviour of social agents. These regularities, created by the agents themselves, are used as informational devices to supplement competitive price information. So Schotter's attempt shares the interest of this chapter in the analysis of organisational forms. It does, however, suffer from a limitation that has been noted in other contexts—it is basically concerned with dynamic adjustments to a static state of nature.[55] Can analysis be widened to deal with situations in which the strategies and payoffs are shaped by the form of organisation?

The information economy

All these matters might perhaps continue to be neglected if information and organisation costs were minor. However, the use of a revised national accounting approach, supplemented by occupational classifications, has tended to confirm the increasing information intensity of economic systems. A recent OECD publication is a convenient way of summarising this approach and its results (see Table 8.1 and Figure 8.1).[56] Figure 8.1 shows that all the countries covered have

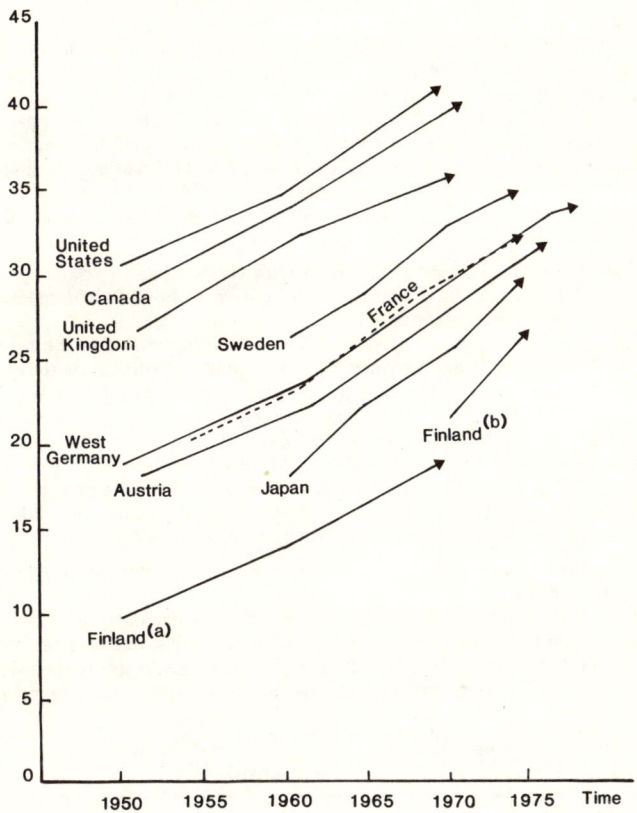

*Data from Finland was derived from two separate sources: (a) Pietarinen; (b) the Central Statistical Office of Finland, both sources using a rather more restricted definition of information occupations than other countries. Absolute values for any given year are, therefore, not strictly comparable with other countries, although the trend is still of interest.
  Source: OECD (1981), *Information Activities, Electronics and Telecommunications Technologies: Impact on Employment, Growth and Trade*, Paris, Vol. I, p. 25.

Figure 8.1. Information occupations as a percentage of economically active.

Table 8.1. Typology of information occupations

Information producers:	scientific and technical workers (components)
	market search and co-ordination specialists
	information gatherers
	consultative services
Information processors:	administrative and managerial
	process control and supervisory
	clerical and related (components)
Information distributors:	educators
	communication workers
Information infrastructure occupations:	information machine workers
	postal and telecommunications

Information producers create new information or package existing information into a form appropriate to a particular recipient.

'Scientific and technical' workers are primarily engaged in research, development and other inventive activities. 'Information gatherers' includes a variety of occupations which, by investigation and assessment, are mainly concerned with creating new information. 'Market search and co-ordination specialists' principally provide, via search activities, market information to buyers, sellers, or (as in brokerage) both. Finally, 'consultative services' are primarily engaged in applying a pre-existent body of information to the particular needs of the 'client' or 'situation'.

Information processors are primarily concerned with receiving and responding to information inputs. The response may be to decide, to administer, or to perform some manipulative operation upon the information inputs.

'Administrative and managerial' occupations receive information in the form of detail about firm (or departmental) performance and environment, instructions from above, and so on, all of which are processed into some form of communication to those superior or subordinate to them. Their job is to decide, organise, plan, interpret or execute policy, whether in private or public undertakings. 'Process control and supervisory' occupations also co-ordinate and control though usually in the more specific context of a particular technical process or body of subordinate workers engaged upon such a process. 'Clerical and related' occupations receive information inputs as correspondence and data, verbal or recorded, and manipulate such inputs into a form appropriate to the employer.

Information distributors are primarily concerned with conveying information from the initiator to the recipient.

'Educators' mainly convey information which has already been produced and 'communication workers' include a variety of occupations in the news and entertainment media. Both groups include elements of information 'production' (as, respectively, with research activities of university teachers and investigative journalism), but the primary activity is considered 'distributive'.

Information infrastructure occupations install, operate and repair the machines and technologies used to support the previous information activities.

Source: OECD (1981), *Information Activities, Electronics and Telecommunications Technologies: Impact on Employment, Growth and Trade*, Paris, Vol. I, p. 24.

undergone a profound change in their occupational structure, the share of information occupations in the economically active increasing by about 3 per cent in each five-year period. Such occupations account for more than one-third of the labour force in these countries. About two in every three people in these occupations are engaged in routine information-handling activities.

There are no doubt many problems in the collection of such statistics. Nevertheless, the analysis has involved several established dichotomies: for example, information vs. non-information, information investment vs. information consumption, routine vs. non-routine information activities. The question to be raised here is whether this approach can incorporate a treatment of organisational change. It is possible, for example, to separate out the creation of new knowledge and then treat the balance of information activity as organisational —as the cost of running the system. This procedure has several deficiencies.

If R & D, as now conventionally measured, is taken as the expenditure on new knowledge creation, the net is too narrow in that this scientist's definition of R & D excludes a significant part of innovation costs that could properly be held to create knowledge. At the same time, much expenditure that is little more than the maintenance of stocks of knowledge and manpower will be included. More significantly in the present context, the residual information expenditures after the exclusion of R & D will include expenditures both to create and maintain organisational capital. These dividing lines are blurred by the difficulties of giving precise meanings to the obsolescence of information and organisational capacity.

The possibilities of using the distinction between production and organisation, the latter being equated with an adjusted information sector, have been touched on by Jonscher.[57] Of course, it is easy to overlook the possibility that expenditures fail to achieve the objectives that led to their being incurred. Because information technologies seem so difficult to evaluate, this form of failure has been labelled 'the xerox effect'.[58] The further possibility that activities included in the information sector may be unproductive is emphasised by Bhagwati:[59] for example, lobbying for protection, competing for a share of licences, and evading governmental regulations. It should, however, be noted that these activities may draw upon the same organisational capacities as other activities deemed of social value in a dynamic context, for example, innovation.

Organisation and information structures[60]

In 1972 Arrow was reported as seeing information economics 'as an effort to plug a hole' in the theory of general equilibrium.[61] Even if no more than this were involved, his position represented a major advance on the view that information could be treated as just one more commodity. (Perhaps some of the inadequacies of this view can be probed by considering what is implied by the assumption of a perfect holiday or perfect transport or perfect manufactured goods!) Simon has pointed out that there are just 'some standard techniques for avoiding a separate theory of procedural rationality—in particular, the proposal that we simply fold in the costs of computation with all of the other costs in the general optimisation problem'.[62] He believed that this would leave 'unanswered all of the fascinating and important questions of what constitutes an efficient decisionmaking procedure'.[63] Information economics addresses these questions. This chapter has so far endeavoured to interpret its emergence, to provide a highly selective guide to some of its literature, and to comment upon significant research direactions. Perhaps echoing Chamberlin's well-known paper, an alternative title for the preceding section of this chapter would have been 'The Organisation as a Variable'. Economists may have to draw upon organisational science and information science in this effort,[64] but history[65] will be a necessary companion discipline.[66]

The organisation is an information mechanism, a mechanism shaped by the balance between perceived information costs and benefits, but with an inbuilt tendency to consider the familiar rather than the novel, the immediate rather than the remote, the cautious rather than the adventurous. As Arrow argues, the organisation is an information mechanism that can achieve important gains from specialisation. There are, however, consequential costs that have to be set against those gains, namely, costs of co-ordination and costs of transmission. The efficiency of internal communication channels can be improved by a choice of code: 'all the known ways, whether inscribed in formal rules, for conveying information'.[67] The economic effects of such coding work out in several ways. First, it can reduce, although it will not eliminate, the tendency for communication costs to rise as the size of the organisational unit increases. Secondly, 'it creates an intrinsic irreversible capital commitment of the organisation'.[68] The more complex the code, the more significant the capital nature of the information mechanisms. The organisation comes to have a distinct identity and its employees to be increasingly skilled in, but limited to, the code of their organisation.

If all firms had the same code, it would be transferable. Why is there not a strong tendency to move towards such a state of affairs? Arrow suggests there are several reasons. There may be many optimal codes; it might be added that where information is not a line item, little is done to evaluate the performance of the coding system. A second major consideration is that the code tends to be selected and shaped according to the expectations that existed when the organisation was created. Because of their capital nature, codes will be modified only slowly over time. The result is that, as Arrow argues, the combination of uncertainty, indivisibility, and capital intensity associated with information channels mean that history matters a great deal and that the very pursuit of efficiency may lead to rigidity and unresponsiveness to further change. Put in a different way, it might be said that organisations can become locked into their information structures just as readily as into their holdings of other assets such as machines and buildings. Several implications are worth noting. First, 'shortrun efficiency and even flexibility within a narrow framework of alternatives may be less important in the long run than a wide compass of potential activities'.[69] Wishful thinking about the value of information may be far less important in countering organisational obsolescence than turnover of decision-makers. Secondly, great care needs to be exercised in adding functions to existing organisations. A classic illustration is that of adding management control to accounting and budgetary departments. 'New Organisations for New Tasks' should always receive consideration.

'Underutilisation' of information

These organisational aspects have an important bearing upon the alleged widespread 'underutilisation' of information resources. Many research institutes and information services hope to reach and indeed aver that they can reach a much higher proportion of their 'client' population than they do at present. It might be expected that such an effort would encounter rising marginal cost, and so would come into conflict with policies directed to maximisation of fee income where those are in operation. However, for the moment there are other questions to be considered. Has a firm been 'reached' when a technical paper lodges in its bookshelves or is dispatched to its wastepaper basket? Has a firm been 'reached' when, influenced by the authoritative arguments of a prestigious institute, it adopts a new piece of hardware and yet fails to make the consequential adjustments in staff, financing, marketing strategy, and the like needed to realise productivity gains? Has a firm been 'reached' when it accepts a 'free' service but fails to evaluate that service simply because it is 'free' and not integrated into normal operations? These questions point to a few of the many problems linked to 'underutilisation' or ineffective utilisation of information resources. Comparable problems characterise the experience of many types of service. Librarians recognise that they reach only a small proportion of the community and prefer to speak of potential rather than non-users. UNISIST-type ventures at the global level tend to be built upon faith in the creation of universal access without looking at the problems at the workface. Many information service institutions seek to expand for reasons inherent in their own structures without involvement in the costly process of ensuring that their service is fully and efficiently enmeshed with the overall functioning of the client enterprise.

To begin to explore this 'underutilisation' phenomenon, use will be made of a set of stylised facts and a rather abstract economic model.[70] The stylised facts are:

(a) low willingness to pay by potential and actual users;
(b) high elasticity of demand;
(c) low volume of service actually demanded; and
(d) high value of the information provided.

The first important point to note is the contrast between the weak or underdeveloped state of the market and the alleged high value of the information preferred. 'It is important to recognise that it is the value of the service, not the value of the information, that determines the economic viability' of the provider of the information.[71] If the provider of the information is in a complete monopoly position, then the value of the service can approach the value of the information. If there are competing suppliers, the value of the service will be limited to the cost of the best alternative information source. Assuming that all the alternatives are clearly perceived and can be evaluated, it is possible to say that the offering of an information service has two initial effects on the individual user. It can lead that user to substitute the new service for old services, which can, of course, include in-house provision, and it can induce expanded demand because it makes possible an increased benefit from the use of information.

Even in these assumed conditions, the structure of the 'industry' providing the information service is clearly of great importance. In the information economy, this industry has grown. Well over half a century ago Knight recognised its importance. One wonders what he would have thought of the shifting and overlapping boundaries of the modern information market place in the Oettinger-Campaigne Harvard mappings.[72] To these marketplace contributions must be added the rich variety of state-provided services. Is not competition then assured on the supply side? For a variety of reasons, caution should be exercised before an affirmative reply. Most importantly, this is a 'market' characterised by rapid, dynamic change and a significant element of monopoly power may derive from a temporary state of affairs. While conventional economic analysis may lead to the happy conclusion that this element of monopoly will be eroded, that is not much consolation if the life of the information product is short.

There will, of course, be parallel considerations on the demand side. The 'information-as-a-public-good' debate implicitly assumes that there is a public or at least a significant number of potential users with both capacity and willingness to use the information. In contrast, in many economies the number of potential users of technical information, for example, patent information on pharmaceuticals, may be small indeed. Where this monopsonistic element enters, the conclusion must accordingly be modified. Quite clearly an element of indeterminacy will be introduced if, as is likely, the situation on either side tends to be oligopolistic.

A second feature on the demand side arises with the aggregation of the demands of individual users. In the textbook case of a competitive market, simple aggregation is permissible. There may even be a good reason for thinking that this structure fosters diffusion. In those cases where there are many producers—even if the competititve conditions are not otherwise met—the interests of the individual user of the information are not threatened by acquisition and use of the same information by other members of the producing group.

This is often the case in agriculture. In fact, it is quite possible that wider diffusion of an innovation might reduce costs to all, for example, through economies of scale and design improvement in the production of equipment. A similar possibility of mutual gain that can lead to the incurring of private costs to disseminate information has been noted by Hirshleifer and Riley:

> An oil firm that has developed a new method of deep recovery might, for example, reap a speculative payoff by buying up options on tracts whose petroleum now lies too deep to be recovered. One important implication of the speculative reward of invention is that it motivates the possessor of information to disseminate it widely and even gratuitously—after having made his speculative commitment.[73]

While those responsible for information services to industry might wish that business firms would behave like farmers, they must reckon with the relevance of industrial structure. Where the competitive conditions are not fulfilled, firms clearly have a lessened incentive to purchase the information. Other circumstances can act as reinforcement. If the information can be purchased freely by all comers, there is always the possibility of waiting until the method, equipment, or product has been tried and possibly improved before copying, assuming that patent barriers can do no more than impose a time and cost penalty to be set against the disadvantages that might well have been faced by the pioneer. If the firm is multiproduct, it has a considerable armoury on which to draw for its response to new technological competition.

So far this discussion of 'underutilisation' of information has been concerned with well-defined optimality problems. The appropriateness of the user organisation has not been involved. The concept of a user of information services calls for careful thought. The ways in which the provision of an information service must, in order to be fully effective, complement the organisational conditions and decision processes within the individual organisation are not standardised. One might argue for a concept of 'targeted' information. Consideration of organisation-specific circumstances is required in addition to taking into account the market structure conditions. Finally, it is as well to remember that examination of alternative models, such as the Nelson and Winter variety,[74] tend to leave the impression that the crucial condition is the separation of learning and doing. However, revision of the Nelson and Winter model in terms of the capital nature of the costs of information implies that rigidity and unresponsiveness can develop even when the organisation has been shaped in a climate initially permitting separate R & D.

Conclusion

This chapter has attempted to show that the emerging information economics is contributing to the understanding of the process of technological change.

Information is a resource and much of information economics is simply an elucidation of the implications of that statement, in opposition to traditional modes of thought that have neglected the role of information. There is, however, a further and even more important contribution: it is no longer possible to treat the institutional framework with its organisational components as given. Economics must turn its attention to the optimal design and use of organisation as well as information resources.

There are profound implications for economic theory. In statics many widely accepted conclusions can be reversed by a simple, even slight, change in the assumptions made about the role of information. When attention shifts to the dynamics of growth, information economics would appear to favour an evolutionary modelling, to raise real doubts about the superiority of any given system of organisation, and to increase the burden of the policy-makers, now denied the strong conclusions that have, erroneously, been carried from static models to the real world.[75]

Notes and references

1. This section is a modified version of the author's paper of the same title presented at the American Economic Association conference in New York, December, 1982.
2. See Knight, F. (1921), *Risk, Uncertainty, and Profit*, Boston, Houghton Mifflin; Hayek, F. von (1937), 'Economics and knowledge', *Economica*, new series, 4: 33-54; Hayek, F. von (1975), 'The pretence of knowledge', *Swedish Journal of Economics*, 77 (4): 433-42; Marschak, J. (1974), *Economic Information, Decision and Prediction. Selected Essays: Vol. II: Economics of Information and Organization*, Boston, D. Reidel; Bell, D. (1973), *The Coming of Post-Industrial Society*, New York, Basic Books; Bell, D. (1979), 'The information society' in Dertouzos, M. and J. Moses (eds), *The Computer Age: A Twenty-Year View*, Cambridge, Mass., M.I.T. Press, pp. 163-211; Machlup, F. (1963), *The Production and Distribution of Knowledge in the United States*, Princeton, N.J., Princeton University Press; Machlup, F. (1980), *Knowledge: Its Creation, Distribution, and Economic Significance. Vol. I: Knowledge and Knowledge Production*, Princeton, N.J., Princeton University Press; Machlup, F. (1982), *Knowledge: Its Creation, Distribution, and Economic Significance. Vol. II: The Branches of Learning, Information Sciences, and Human Capital*, Princeton, N.J., Princeton University Press; Shackle, G. (1969), *Decision Order and Time in Human Affairs*, Cambridge, Cambridge University Press, 2nd edn; Boulding, K. (1966), 'The economics of knowledge and the knowledge of economics', *American Economic Review*, 56 (2): 1-13; Kornai, J. (1971), *Anti-Equilibrium*, Amsterdam, North-Holland; Simon, H. (1978), 'On how to decide what to do', *Bell Journal of Economics*, 9 (2): 494-507; Arrow, K. (1974), *The Limits of Organization*, New York, Norton; and Arrow, K. (1979), 'The economics of information' in Dertouzos, M. and J. Moses (eds), *The Computer Age: A Twenty-Year View*, Cambridge, Mass., M.I.T. Press, pp. 306-17.
3. For example, Spence, A. M. (1974), 'An economist's view of information' in Cuadra, C. and A. Luke (eds), *Annual Review of Information Science and Technology*, Washington, D.C., American Society for Information and Science, Vol. 9; Tomasini, L. (1974), 'The economics of information: a survey', *Economie Appliquée*, 27 (2/3): 319-37; Lamberton, D. (ed.) (1971), *The Economics of Information and Knowledge*, Harmondsworth, Penguin; Lamberton, D. (ed.) (1974), *The Information Revolution*, American Academy of Political and Social Science, *Annals*, 412; Lamberton D. (1975), *Who Owns the Unexpected?*, St Lucia, University of Queensland Press; Lamberton, D.

(1978), 'The economics of communication' in Rahim, S. *et al., Planning Methods, Models, and Organizations*, Honolulu, East-West Communication Institute; Lamberton, D. (1982), 'The theoretical implications of measuring the communication sector' in Jussawalla, M. and D. Lamberton (eds), *Communication Economics and Development*, New York, Pergamon, pp. 36-59; and Hirshleifer, J. and J. Riley (1979), 'The analytics of uncertainty and information—an expository survey', *Journal of Economic Literature,* 17: 1375-421.

4. See American Economic Association Conference (1966), *American Economic Review,* 56 (2); Princeton Symposium: The Economics of Information (1976), *Quarterly Journal of Economics,* 90 (4); Lund Symposium on The Economics of Information (1974), *Swedish Journal of Economics,* 76; Bell Journal Symposium on the Economics of Internal Organization (1975), *Bell Journal of Economics,* 6 (1); Perlman, M. (ed.) (1977), *The Organization and Retrieval of Economic Knowledge*, London, Macmillan; Stanford Symposium on Economics of Information (1977), *Review of Economic Studies,* 44: 389-599; Galatin, M. and R. Leiter (eds) (1982), *Economics of Information*, Boston, Martinus Nijhoff; UNESCO special issue on *Economics of Information and Information for Economists* (1976), *International Social Science Journal,* 28 (3); and Goldberg, R. and H. Lorin (eds) (1982), *The Economics of Information Processing*, 2 vols, New York, Wiley-Interscience.

5. The following are illustrative:
   000 Grossman, S. and J. Stiglitz (1980), 'On the impossibility of informationally efficient markets', *American Economic Review,* 70 (3): 393-408; Welch, R. (1980), 'Vertical and horizontal communication in economic processes', *Review of Economic Studies,* 47: 733-46.
   100 Jussawalla, M. and D. Lamberton (eds) (1982), *Communication Economics and Development*, New York, Pergamon; Gertler, M. (1982), 'Imperfect information and wage inertia in the business cycle', *Journal of Political Economy,* 90 (5): 967-87.
   200 OECD (1981), *Information Activities, Electronics and Telecommunications Technologies: Impact on Employment, Growth and Trade*, 2 vols, Paris; Jussawalla, M. and Chee-Wah Cheah (forthcoming), *Towards an Information Economy: A Case Study of Singapore*.
   300 Mansfield, E. (1982), 'Tax policy and innovation', *Science,* 215 (4538): 1365-71: Pommerehne, W. and F. Schneider (1978), 'Fiscal illusion, political institutions, and local public spending', *Kyklos,* 31 (3): 381-408.
   400 Caves, R., H. Crookell and J. Killing (1982), 'The imperfect market for technology licenses', Discussion Paper No. 903, Harvard Institute of Economic Research; NTIA (National Telecommunications and Information Administration) (1981), *Trade Issues in Telecommunications Information*, 4 vols, Washington, D.C., US Department of Commerce.
   500 Hodgson, J. and D. Eade (eds) (1982), *Information Systems in Public Administration*, Amsterdam, North-Holland; McMillan, J. (1982), 'Robotics. Will the corporation be managed by machines?', *Cost and Management,* 56 (4): 2-7.
   600 Bradburd, R. and A. Over (1982), 'Organizational costs, "sticky equilibria", and critical levels of concentration', *Review of Economics and Statistics,* 64: 50-8; Wigand, R. (1982), 'The communication industry in economic integration: the case of West Germany', *Social Networks,* 4 (1): 47-79.
   700 Riemenschneider, C. and J. Bonnen (1979), 'National agricultural information systems: design and assessment: in Blackie, M. and J. Dent (eds), *Information Systems for Agriculture*, London, Applied Science Publishers; Crouch, B. and S. Chamala (eds) (1981), *Extension Education and Rural Development, International Experience in Communication and Innovation*, New York, Wiley, Vol. 1.
   800 Backhouse, R. (1982), *Information Services for Trade Unionists*, London, Elm Publications; Mauro, K. (1982), 'Strikes as a result of imperfect information', *Industrial and Labor Relations Review,* 35 (4): 522-38.
   900 Beales, H., R. Craswell and S. Salop (1981), 'The efficient regulation of consumer

information', *Journal of Law and Economics*, **24**: 491–544; Schuman, P. (1982), 'Information justice', *Library Journal*, **107** (11): 1060–66.
6. For example, Porat, M. (1977), *The Information Economy*, 9 vols, Washington D.C., US Department of Commerce, Office of Telecommunications; Sowell, T. (1980), *Knowledge and Decisions*, New York, Basic Books; McCall, J. (ed.) (1982), *The Economics of Information and Uncertainty*, Chicago, University of Chicago Press; Mackaay, E. (1982), *Economics of Information and Law*, Hingham, Kluwer Boston; and Nelson, R. and S. Winter (1982), *An Evolutionary Theory of Economic Change*, Cambridge, Mass., Harvard University Press.
7. Coase, R. (1977), 'Economics and contiguous disciplines' in Perlman, M. (ed.), *The Organization and Retrieval of Economic Knowledge*, London, Macmillan, p. 484.
8. Lamberton, D. (1976), 'National policy for economic information', *International Social Science Journal*, **28** (3): 449–65.
9. OECD (1971), *Information for a Changing Society. Some Policy Considerations*, Paris, p. 13.
10. For example, NCLIS (National Commission on Libraries and Information Science) (1976), *National Information Policy*, Washington, D.C., US Government Printing Office; Dunn, D. (1982), 'Developing information policy', *Telecommunications Policy*, **6** (1): 21–38; and Bjørn-Andersen, N., M. Earl, O. Holst and E. Mumford (eds) (1982), *Information Society: For Richer, For Poorer*, Amsterdam, North-Holland.
11. Dunn, D. (1982), 'Developing information policy', *Telecommunications Policy*, **6** (1): 38.
12. JIPDEC (Japan Information Processing Development Center) (1981), *Preliminary Report on Study and Research on Fifth-Generation Computers 1979-80*, p. 86.
13. Some might say that the final badge is having its own journal, the publication of *Information Economics and Policy* having been announced in 1982 by North-Holland Publishing Company. While such a publication fits the general pattern of increasing specialisation in information activities, one might have mixed feelings because some colleagues are inclined to interpret information economics as just another field of specialisation rather than an attempt to address the really big problems of economics.
14. Robbins, Lord (1968), *The Theory of Economic Development in the History of Economic Thought*, London, Macmillan.
15. Cannan, E. (1929), *Review of Economic Theory*, London, Frank Cass, pp. 122-5.
16. Ibid., p. 83
17. Lamberton, D. (1974), 'Dugald Stewart's work on technological change', Department of Economics, University of Queensland, mimeo.
18. Hodgskin, T. (1827), *Popular Political Economy*, New York, Augustus M. Kelley Reprints of Economic Classics (1966), p. 53.
19. Williams, B. (1964), 'Review of F. Machlup, *The Production and Distribution of Knowledge in the United States*', *Economic Journal*, **74** (293): 174.
20. Knight, F. (1921), *Risk, Uncertainty, and Profit*, Boston, Houghton Mifflin, p. 261.
21. Ibid., p. 260.
22. Idem.
23. See Welch, R. (1980), 'Vertical and horizontal communication in economic processes', *Review of Economic Studies*, **47**: 733–46.
24. Spence, A. M. (1974), 'An economist's view of information' in Cuadra, C. and E. Luke (eds), *Annual Review of Information Science and Technology*, Washington D.C. American Society for Information Science, Vol. 9. p. 57.
25. Ibid., p. 58.
26. Stiglitz, J. (1977), Stanford symposium on Economics of Information, *Review of Economic Studies*, **44**: 389.
27. Idem.
28. Idem.
29. Hirshleifer, J. and J. Riley (1979), 'The analytics of uncertainty and information— an expository survey', *Journal of Economic Literature*, **17**: 1393.
30. Ibid., 1397.
31. Ibid., 1405.

32. Ibid., 1414.
33. Arrow, K. (1979), 'The economics of information' in Dertouzos, M. and J. Moses (eds), *The Computer Age: A Twenty-Year View*, Cambridge, Mass., M.I.T. Press, p. 300.
34. Ibid., p. 310.
35. Idem.
36. Arrow, K. (1974), *The Limits of Organization*, New York, Norton, p. 49.
37. Starrett, D. (1976), 'Social institutions, imperfect information, and the distribution of income', *Quarterly Journal of Economics,* 90 (2): 282.
38. Begg, D. (1981), *The Rational Expectations Revolution in Macroeconomics: Theories and Evidence*, Oxford, Philip Allan.
39. O'Driscoll, G. (1979), 'Rational expectations, politics, and stagflation' in Rizzo, M. (ed.), *Time, Uncertainty and Disequilibrium*, Lexington, Mass., Lexington Books, p. 157.
40. Simon, H. (1978), 'On how to decide what to do', *Bell Journal of Economics,* 9 (2): 494-507.
41. Hahn, F. (1973), *On the Notion of Equilibrium in Economics*, Cambridge, Cambridge University Press, p. 40.
42. Ibid., p. 504.
43. Ibid., p. 505.
44. Idem.
45. Bradburd, R. and A. Over (1982), 'Organizational costs, "sticky equilibria", and critical levels of concentration', *Review of Economics and Statistics,* 64: 51.
46. See p. 56 in this book.
47. This is the title of Ch. IV of Andrews, P. (1949), *Manufacturing Business*, London, Macmillan.
48. For example, Wolfe, A. (1977), 'The supranational organization of production: an evolutionary perspective', *Current Anthropology,* 18 (4): 615-35.
49. Stout, D. (1980), 'The impact of technology on economic growth in the 1980s', *Daedalus,* 109 (1): 165.
50. Tinbergen, J. (1978), 'Alternative forms of international co-operation: comparing their efficiency', *International Social Science Journal,* 30 (2): 223-37.
51. 'For the study of the individual firm it must be recognized that the firm possesses a stock of capital, an organization, and a stock of information which will be modified by the firm's own endeavours and not merely as an automatic result of the operation of the market', Lamberton, D. (1965), *The Theory of Profit*, Oxford, Basil Blackwell, p. 7.
52. McCain's 'tradition' comes close to Teubal's concept. See McCain, R. (1981), 'Tradition and innovation: some economics of the creative arts, science, scholarship, and technical development' in Galatin, M. and R. Leiter (eds), *Economics of Information*, Boston, Martinus Nijhoff, pp. 173-208.
53. Cf. Nelson, R. (1980), 'Production sets, technological knowledge, and R & D: fragile and overworked constructs for analysis of productivity growth?', *American Economic Review,* 70 (2): 62-7.
54. Schotter, A. (1981), *The Economic Theory of Social Institutions*, Cambridge, Cambridge University Press.
55. Cf. Slater, J. (1982), 'Review of Schotter, *The Economic Theory of Social Institutions*', *Economic Journal,* 92 (367): 714-15.
56. OECD, (1981), *Information Activities, Electronics and Telecommunications Technologies: Impact on Employment, Growth and Trade*, 2 vols, Paris.
57. Jonscher, C. (1982), 'Notes on communication and economic theory' in Jussawalla, M. and D. Lamberton (eds) *Communication Economics and Development*, New York, Pergamon, pp. 60-9.
58. Lamberton, D., S. Macdonald and T. Mandeville (1982), 'Productivity and technological change', *Canberra Bulletin of Public Administration,* 9 (2): 23-30.
59. Bhagwati, J. (1982), 'Directly unproductive, profit-seeking (DUP) activities', *Journal of Political Economy,* 90 (5): 988-1002.
60. This section draws upon the author's consultant report, *Policies and Programmes in*

Stimulation of Innovation and Information Flow, prepared for the Institute for Industrial Research and Standards, Dublin 1982; and a paper, 'Development and information structures', presented at the Pacific Science Congress, Dunedin, New Zealand, February 1983.
61. *New York Times* (1972), 'Nobel winner engrossed by balancing act', 26 November, F5.
62. Simon, H. (1978), 'On how to decide what to do', *Bell Journal of Economics,* 9 (2): 506.
63. Idem.
64. For example, Ramos, A. (1982), *The New Science of Organizations: A Reconceptualization of the Wealth of Nations*, Toronto, University of Toronto Press.
65. For example, Chandler, A. D. (1977), *The Visible Hand: The Managerial Revolution in American Business*, The Belknap Press of Harvard University Press; Williamson, O. (1981), 'The modern corporation: origins, evolution, attributes', *Journal of Economic Literature,* 19 (4): 1537–68.
66. Some interesting cases of and suggestions for organisational innovation might be Japan's MITI; 'Commerce on the campus' (*The Economist* (1982), 23 January, p. 76); a high-technology Morrill Act (Botkin, J. et al. (1982), 'High technology, higher education, and high anxiety', *Technology Review,* 85 (7): 49-57); Los Angeles' Telacu and Uno ventures (*The Economist* (1982), 'American survey: Los Angeles', 3 April, pp. 57–88); and Australia's Snowy Mountains Engineering Corporation (Price, D. and J. Elston (1980), 'Developing countries' avenues for assistance—advantages and disadvantages', *First International Conference on Technology for Development 1980*, Canberra, Institution of Engineers, Australia).
67. Arrow, K. (1974), *The Limits of Organization*, New York, Norton, p. 55.
68. Idem.
69. Ibid., p. 59.
70. Sassone, P. (1981), 'A theory of market demand for information analysis center services' in Mason, R. and J. Creps (eds), *Information Services: Economics, Management, and Technology*, Boulder, Colorada, Westview Press/Praeger, pp. 23–38.
71. Ibid., pp. 24-5.
72. See Oettinger, A. (1980), 'Information resources: knowledge and power in the 21st century', *Science,* 209: 191-8; Compaigne, B. (1981), 'Shifting boundaries in the information market place', *Journal of Communication,* 31 (1): 132-42.
73. Hirshleifer, J. and J. Riley (1979), 'The analytics of uncertainty and information—an expository survey', *Journal of Economic Literature,* 17: 1405.
74. Nelson, R. and Winter S. (1982), *An Evolutionary Theory of Economic Change*, Cambridge, Mass., Harvard University Press. For a summary statement of the relevance to economic theory, see Nelson, R. (1980), 'Production sets, technological knowledge, and R & D: fragile and overworked constructs for analysis of productivity growth?', *American Economic Review,* 70 (2): 62-7.
75. Cf. Mandeville, T., S. Macdonald and D. Lamberton (1980), 'The fortune tellers' new clothes: a critical appraisal of IMPACT's technological change projections to 1990/91', *Search,* 11 (1/2): 14-17.

PART III
DIFFUSION, TECHNOLOGY TRANSFER AND TRADE

9. Theoretical approaches to the analysis of the diffusion of new technology
P. Stoneman

Introduction

Consider that an innovation of a new technology has already occurred. For the new technology to have any considerable impact on the economy the innovation must be emulated by other actors in the economy. This process of emulation is the process of diffusion. For the sake of brevity in this chapter, only the diffusion of a new process technology within a given industry is considered. The basic measure of diffusion will be the proportion of the industry's output produced with the new technology. Two main themes will underly the presentation: (i) that the explanation of the diffusion process must be based on rational behaviour, and (ii) that the observed diffusion pattern will be the result of the interaction between demanders and suppliers of the new technology. It is not the case that the main body of the diffusion literature has always stayed true to these themes, but they seem to represent a good starting point. The main objectives of this chapter are to illustrate why, in the majority of studies, a plot of the extent of diffusion against time is sigmoid (S-shaped), and what factors determine the speed of diffusion.

Two recent surveys of the diffusion literature have been published by Davies and David, so it is not necessary to undertake a complete coverage here.[1] This chapter will concentrate on three main approaches on the demand side whereby the diffusion pattern is the result of: (a) information and learning, (b) differences between firms, or (c) the strategic behaviour of firms. On the supply side: (a) optimal pricing policy, (b) learning by doing economies, and (c) market structure are considered.

Learning processes and diffusion

Models of diffusion based on learning by potential users of a new technology have a long history. Early attempts to model this process likened the spread of new technology to the spread of a disease.[2] In these models a new technology is adopted when its existence is known, and this knowledge is passed on by interpersonal contact. Mansfield's seminal contribution refines the approach considerably.[3] Mansfield argues that at a moment in time the factors that affect the rate at which non-users take up a new technology include the profit to be gained from use and the riskiness attached to use. It is argued that profitability remains constant over the diffusion process, but as the level of use increases entrepreneurs learn more about the new technology, reduce their estimate of riskiness, and this leads to further use. By essentially assuming that functions are of the appropriate form, it is possible to argue that diffusion, appropriately defined, can be summarised by a logistic curve. Specifically, Mansfield shows that if we define m_{jt} = number of users of technology j in time period t, and $n_j = \lim_{t \to \infty} m_{jt}$ then

$$\frac{m_{jt}}{n_j} = \frac{1}{1 + \exp(-a_j - \beta_j t)}, \quad (1)$$

where

$$a_j = \log\left(\frac{m_{j0}/n_j}{1 - m_{j0}/n_j}\right),$$

and β_j is a measure of the 'inter-firm' diffusion speed. It is also shown by the same procedure that if we define α_{it} as the proportion of firm i's output in time t produced with a new technology then

$$\alpha_{it} = \frac{1}{1 + \exp(-\gamma_i - b_i t)}, \quad (2)$$

where

$$\gamma_i = \log\left(\frac{\alpha_{i0}}{1 - \alpha_{i0}}\right),$$

and b_i is a measure of the intra-firm diffusion speed. It is argued that both β_j and b_i will be linear functions of, among other variables, the constant level of profitability. Thus the more profitable a new technology, the faster will be the rate of inter-firm diffusion and the more profitable is a new technology to firm i, the faster will be the firm's rate of intra-firm diffusion. Obviously with models for determining α_{it} and m_{jt}/n_j one can obtain results on the proportion of industry output produced on the new technology.

This framework, however, considers only the demand side of the diffusion process. To determine fully the diffusion path one must consider the supply side as well. Glaister has considered the addition of a supply side to the epidemic and Mansfield-type models.[4] He represents the number of users in time t by an alternative version of equation (1),

$$\frac{dm_t}{dt} = \beta(p_t)(n - m_t)m_t, \quad (3)$$

in which the j subscript has been dropped and β is now a function of the price, p, that users have to pay for the goods embodying new technology. For a strong link to the Mansfield model one might consider p to proxy profitability. If p is constant over time, then $\beta(p_t)$ will be constant and (3) yields the logistic curve predicted by Mansfield. Glaister then allows that in each time period a user will buy $h(p_t)$ units of the new good, which one might think of as representing the intra-firm diffusion process. If q_t is allowed to be the output of the goods embodying the new technology in time t, then for demand and supply to be equal it is necessary that (4) holds

$$q_t = m_t h(p_t). \tag{4}$$

The supplying industry is assumed to be monopolised, with the single producer maximising the present value of the firm, V, given by (5)

$$V = \int_0^\infty e^{-rt}\{p_t q_t - C(q_t)\}dt, \tag{5}$$

where $C(q_t)$ is the cost of producing the new goods.

In an example involving constant returns to scale, a constant elasticity function $h(p_t)$, and a constant elasticity function $\beta(p_t)$, Glaister shows that the optimal pricing policy involves a price below cost initially with price rising above cost as time proceeds, with price converging on a long-run equilibrium above cost. The changing price changes β. The net result is that the growth curve of the stock of the new good is still sigmoid, but not in fact logistic; it has a positive skew.

Glaister takes his work further by also considering advertising policy. Holding price constant, he shows that the optimal policy is to have an early splurge of advertising, which then dies out. The net result is that the growth curve of ownership has a positive skew again. This is of a particular interest, for as the epidemic results are so dependent on information transfer, it seems that an obvious extension is to consider advertising as a means of information transfer.

This model of Glaister, although interesting, is deficient on two main counts: (i) no allowance is made for oligopolistic rivalry or entry into capital goods supply, and (ii) there is no learning in capital goods production. Both deficiencies have been approached in a different context by Spence, which is considered later.[5] What is more useful here is to consider Metcalfe's contribution.[6] Metcalfe assumes that inter- and intra-firm diffusion generate a time path for demand such that

$$\frac{dy}{dt} \cdot \frac{1}{y} = \beta(y_t^* - y_t), \tag{6}$$

where y is demand for the new good in time t. Equation (6) is a way of representing a logistic diffusion curve used by Chow and Stoneman and is justified by Chow on epidemic grounds.[7] In equation (6) y_t^* is 'equilibrium' demand and is allowed by Metcalfe (as by his predecessors) to depend on price, that is

$$y_t^* = c - gp_t. \tag{7}$$

On the supply side Metcalfe assumes that if r_t is the rate of profits in time t, if Π is the fraction of internally generated profits ploughed back in capacity expansion and μ the ratio of external to internal funds invested at any time, then capacity q_t will grow according to (8)

$$\frac{dq_t}{dt} \cdot \frac{1}{q_t} = \Pi(1+\mu)r_t. \tag{8}$$

One should note that in (8), Π and μ are constant by assumption and are not explicitly derived by reference to any maximising or otherwise rational behaviour of the firms. If it is now assumed that there is a fixed coefficient technology for producing the new good with a capital output ratio v and labour output ratio l, with a price for labour w and depreciation at rate δ, then we can state that

$$r = \frac{p_t - w_t l - \delta v}{v}. \tag{9}$$

We now allow that wages increase with labour input so

$$w_t = w_0 + \theta q_t. \tag{10}$$

Then from (8), (9) and (10) we have (11)

$$\frac{dq}{dt} \cdot \frac{1}{q} = \frac{\Pi(1+\mu)}{v} p_t - \frac{(w_0 l + \delta v)}{v} - \frac{\theta l}{v} q_t. \tag{11}$$

If it is allowed that demand and capacity grow in a balanced way so that (12) holds, and $q_t = y_t$ for all t,

$$\frac{dy}{dt} \cdot \frac{1}{y} = \frac{dq}{dt} \cdot \frac{1}{q} \equiv k_t, \tag{12}$$

then from (6), (7), (11) and (12) we generate an expression for k_t, as a function of the parameters of the model and y_t. Metcalfe shows that this function is a logistic curve, with, as the diffusion proceeds, the profitability of adopting changing.

Objection can be raised to this model of supply on the grounds that there is no explicit consideration of the rational behaviour of firms. Spence has considered the capacity decisions of a firm entering a new market and his results do not match Metcalfe's suppositions.[8] The strong point of the analysis, however, is that financial constraints on the supply industry are considered explicitly. One can also object to the analysis because the costs of producing the new good increase over time (see (10)) which is not really supported by the empirical analysis of new products. If some learning economies or further technological advance in the supply industry were introduced, this objection might be overcome.

However, perhaps of more importance is criticism of the epidemic model itself. Davies argues convincingly that to rely on information being spread by contact between individuals as an explanation for diffusion in a modern mass-media society is somewhat unacceptable.[9] Mansfield's version of the epidemic model, though, is more sophisticated. Criticism of this model centres on the rudimentary nature of its choice theoretic framework, and more importantly on its treatment of risk. In Mansfield's model the profit to be derived from changing technology—more correctly, expected profit—remains constant over time. However, risk and/or uncertainty is reducing. It is difficult to conceive of the uncertainty being about any variable other than profitability, but as uncertainty reduces, expected profit by assumption is being held constant. This suggests that reduction in uncertainty is just leading to a confirmation that the initial estimate of profitability was the correct one. This seems to be a strange procedure.

In Stoneman and Lindner *et al.*, there is a reworking of this type of framework.[10] An explicit theory of technique choice is combined with Bayesian learning to generate the demand for new technology. In general the demand curve will not be logistic. The several criticisms detailed above tend to suggest that the epidemic models are not the best way to consider diffusion. We turn then to models that rely on differences among firms.

Probit models

This approach concentrates on the characteristics of the firms in an industry, and as such is not only suitable for generating a diffusion curve, but will also give indications of which firms will be early adopters and which ones late. The principle of these probit models is stated succinctly by David:

> Whenever or wherever some stimulus variate takes on a value exceeding a critical level, the subject of the stimulation responds by instantly determining to adopt the innovation in question. The reason such decisions are not arrived at simultaneously by the entire population of potential adopters lies in the fact that at any given point of time either the 'stimulus variate' or the 'critical level' required to elicit an adoption is described by a distribution of values, and not a unique value appropriate to all members of the population. Hence, at any point in time following the advent of an innovation, the critical response level has been surpassed only in the cases of some among the whole population of potential adopters. Through some exogenous or endogenous process, however, the relative position of stimulus variate and critical response level are altered as time passes, bringing a growing proportion of the population across the 'threshold' into the group of actual users of the innovation.[11]

Let X be the stimulus variate with a relative density function $f(X)$ and $\bar{X}_t$ represent the critical value of the stimulus variate in time t. Then in a non-stochastic version of the model, the proportion of the population who are adopters by time t is given as

$$\int_{\bar{X}_t}^{\infty} f(X_i) \, dX_i.$$

It should be clear that only once one has specified $f(X)$ and $\bar{X}_t$, has this model really any operational significance. David suggests that prior to his contribution there were essentially four classes of models that fitted within this framework:

(a) The entrepreneurial inertia model—in this X is defined as the return necessary to persuade an entrepreneur to change technology.
(b) The information cost model—in this case the returns necessary to induce change are the same for each entrepreneur, but it is considered that entrepreneurs face different search costs.
(c) The fixed capital replacement model—entrepreneurs differ in terms of the gain to be made from new technology for they own capital equipment of different productive capacity.
(d) The vintage-model approach, which is closely related to (c).

Rather than pursue all these models here, just two variants of the probit approach will be discussed. The first will be David's firm size model, and the second Davies' independently produced variant.

David's model defines firm size as the critical variable, and proceeds to discuss how firm size is distributed within an industry and how the threshold value of firm size is determined. It is argued that firm size is lognormally distributed. To determine the critical value of firm size David argues as follows. He considers a capital-embodied process innovation that involves fixed costs above and variable costs below those of the replaced technique. C is defined as the purchase cost of the new equipment (the same for all firms whatever their size) and R as its imputed rental rate (defined as $r(1 - e^{-rd})^{-1}$ where r is the interest rate and d the expected service life of the equipment). On the grounds of simplicity the replaced technique is assumed to be purely labour using. The new technique saves labour input relative to the old technique such that for each unit of output produced the labour saving is L_s. W is the wage rate. As should be obvious, there will be some level of scale (output) at which the labour savings will compensate for the increased capital cost. This defines the critical value of firm size $\bar{X}$.

$\bar{X}$ may be defined as that value of X where (13) holds

$$\bar{X} W L_s = C \cdot R \tag{13}$$

that is,[12]

$$\bar{X} = \frac{1}{L_s} \cdot \frac{C \cdot R}{W} \equiv \frac{1}{\omega} \cdot \frac{1}{L_s}. \tag{14}$$

To generate a diffusion path we need to have $\bar{X}$ or the distribution of X_i changing over time. For the simplest case it is assumed that $f(X_i)$ remains constant over time. David then argues that $\bar{X}$ will change over time as wages rise relative to capital costs, arguing that (15) will hold

$$\frac{d\bar{X}}{dt} = \bar{X}_t \cdot \frac{d\omega}{dt} \cdot \frac{1}{\omega} \equiv \bar{X}_t \cdot \dot{\omega} \tag{15}$$

and that $\dot{\omega} = \lambda$, a constant independent of time, that is, the relative factor prices ω follow an exponential time trend.

Thus defining $D(t)$ as the proportion of the population using the new technology in time t, we have (16)

$$D(t) \equiv Pr\{X_i \geq \bar{X}_t\} = \int_{\bar{X}_t}^{\infty} f(X_i) dX_i, \tag{16}$$

thus

$$\frac{dD(t)}{d\bar{X}_t} = -f(\bar{X}_t), \tag{17}$$

and

$$\frac{dD}{dt} = \frac{dD}{d\bar{X}_t} \cdot \frac{d\bar{X}_t}{dt} = f(\bar{X}_t) \cdot \lambda \cdot \bar{X}_t. \tag{18}$$

David then goes on to show that D, as thus defined (given $f(X_i)$ is lognormal and λ is constant), will trace out the standard cumulative normal curve when plotted against a positive linear transformation of the time variable, which curve is, of course, sigmoid.

Davies' model is similar, but in some ways richer.[13] He argues that because of uncertainty, firms make decisions in a behavioural manner; specifically, a firm i will use a new technology if the expected pay-off period from its use $ER_{it} \leq \bar{R}_{it}$, some critical pay-off period. The expected pay-off period is then a function of firm size X_{it} and other firm characteristics (Y_{ijt}) and time.

$$ER_{1t} = \theta_{1t} X_{it}^{\beta_1} \epsilon_{1it}, \tag{19}$$

where

$$\epsilon_{1it} = \prod_{j=1}^{r} Y_{ijt}^{V_j} > 0,$$

and

$$\theta_{1t} > 0$$

$$\frac{d\theta_{1t}}{dt} \cdot \frac{1}{\theta_{1t}} > 0 \text{ for all } t.$$

$\bar{R}_{it}$ is related to firm size and other characteristics (Z_{ijt}) by

$$\bar{R}_{it} = \theta_{2t} X_{it}^{\beta_2} \epsilon_{2it}, \tag{20}$$

where

$$\epsilon_{2it} = \prod_{j=1}^{u} Z_{ijt}^{\Omega_j} > 0,$$

and

$$\theta_{2t} > 0$$

$$\frac{d\theta_{2t}}{dt} \cdot \frac{1}{\theta_{2t}} > 0 \text{ for all } t.$$

In other words, both the expected pay-off period and the critical period vary with firm size.

Following from the above, it may be stated that $ER_{it} \leq \bar{R}_{it}$ if (21) holds.

$$\frac{\theta_{1t} \cdot X_{it}^{\beta_1} \cdot \epsilon_{1it}}{\theta_{2t} X_{it}^{\beta_2} \epsilon_{2it}} \tag{21}$$

Define

$$\theta_t = \frac{\theta_{1t}}{\theta_{2t}}, \text{ and } \epsilon_{it} = \frac{\epsilon_{1it}}{\epsilon_{2it}} \text{ and } \beta = \beta_1 - \beta_2$$

then (21) may be rewritten as (22).

$$\theta_t \cdot X_{it}^{\beta} \cdot \epsilon_{it} \leq 1. \tag{22}$$

Thus a firm will be using the new technology in a deterministic model if (23) holds.

$$X_{it}^{\beta} \leq (\theta_t \epsilon_{it})^{-1}. \tag{23}$$

Critical firm size is then defined as in (24)

$$\bar{X}_{it} = (\theta_t \epsilon_{it})^{-1/\beta}. \tag{24}$$

Davies now argues that actual firm sizes are lognormally distributed. We thus have a distribution of firm sizes, and a definition of the critical threshold of firm size, so once we specify how critical firm size varies over time, we have a theory of diffusion. For the moment assume ϵ_{it} is constant and the same for all firms. Also for the sake of the argument assume $\beta > 0$. Then the critical value of firm size will change as θ_t changes. Davies considers two types of innovation:

Group A: for which $\theta_t = \alpha t^{\psi}$ $\alpha > 0$,

Group B: for which $\theta_t = \alpha e^{\psi t}$ $0 < \psi < 1$.

He then shows that the proportion of firms using the new technology will follow a cumulative lognormal time path for Group A innovations and a cumulative normal time path for Group B innovations. Group A innovations are relatively cheap and simple innovations that experience major improvements in their early years, but fewer improvements thereafter. Group B innovations are expensive and technically complex, experiencing improvement for many years after their first commercial introduction. Through ϵ_{it} differences between firms can be allowed to influence the diffusion pattern, and having $\beta < 0$ is shown not really to affect the argument. Davies also allows the model to be made more complex by letting the firm size distribution vary and shift over time.

The David and Davies models both predict that the probability of a firm adopting an innovation in time t will be a linear function of the log of firm size. Davies tests this hypothesis and finds strong evidence to support it on a data sample of 22 innovations. He also finds that typically $\beta > 0$, which is also consistent with the David model. Both the David and Davies models require that the profitability of adoption changes over time for a diffusion to be generated. We expect profit to be related to the price paid for the goods embodying the new technology; however, neither author explicitly considers the determination of this price. In Stoneman and Ireland, a supply side is added to provide an analysis explicitly incorporating supply factors.[14] Following the David model, it is argued that the inverse demand function for the stock of goods of the new type x_t, can be written as (25).

$$p_t = g(x_t, t). \tag{25}$$

The supply side in its general form is considered as represented by a present value maximising firm that chooses its output subject to certain constraints. Explicitly the firm maximises V_i, where V_i is given by (26)

$$V_i = \int_0^\infty [p_t q_{it} - c(x_{it}, x_{jt}, q_{it}, t)] e^{-rt} dt, \tag{26}$$

subject to

$$\frac{dx_{it}}{dt} = q_{it}$$

$$\frac{dx_t}{dt} = q_{it} + q_{jt}$$

$$x_0 = 0$$

$$x_t = x_{it} + x_{jt},$$

where subscript j refers to rival firms in the oligopoly game. So, for example, q_{jt} is the output for rivals in time t; x_{it} is the stock of the capital good produced by firm i to time t, and $c(x_{it}, x_{jt}, q_{it}, t)$ represents production costs. The model allows for oligopolistic rivalry, learning by doing and reductions in cost over time because of further technological advance.

The results of the analysis suggest that under a wide range of circumstances sigmoid diffusion will result, but it is not always necessary to have a sigmoid demand relation to ensure this. It is shown that the oligopolistic structure of the supplying industry will affect the speed of diffusion, as will the degree of learning economies and the rate of technological advance in capital goods production. The moral of the analysis is, however, that only by adding an explicitly modelled supply side can the full story of the probit analysis be told. Spence has investigated a similar type of model where learning economies exist, but his model is not explicitly directed at analysing diffusion.[15] He considers learning economies as an entry barrier and these are thus introduced into the analysis.

The game theoretic approach

Reinganum has recently considered a further approach to the analysis of diffusion.[16] She explicitly removes any differences between firms in either characteristics or information, and generates a diffusion from entirely strategic behaviour. An industry of n identical firms is considered. It is assumed that a firm can, given a number of existing users, always make more profit using the new technology than the old, and moreover that the profit will be higher the earlier the date of adoption. However, it is argued that because of adjustment costs, early adoption is more costly. Assuming Cournot behaviour, it is then shown that there exists a Nash equilibrium in which the firms have different adoption dates. Thus a diffusion path exists—that is, although all firms are the same, the equilibrium does not have them all adopting at the same date. It is not shown that this diffusion is sigmoid, nor is there any way of saying which firms adopt early, but the approach is a new and fascinating one. There is no supply side modelling in the framework. The model is also analysed for the effects of changes in n, representing market structure, and it is argued that for a linear demand case, higher n will delay adoption for most users.

The question of the impact of market structure of the user industry on diffusion speed is also analysed by Romeo in the context of the Mansfield model and by Davies in the probit context.[17] Davies finds that a larger number of firms is associated with a lower diffusion speed, as is a lower variance of the log of firm size. Romeo finds that a lower diffusion speed is associated with a higher variance. In Stoneman and Ireland, it is argued that a larger number of firms in the using industry is associated with a higher level of use for all t as is a larger number of producers. In total, therefore, the effect of market structure is somewhat uncertain.

Conclusion

In this chapter we have covered three basic approaches to diffusion, each of which has its own merits. It is argued that the diffusion process is related, on the demand side, to:

(1) learning and uncertainty,
(2) differences between firms,
(3) the nature of the oligopoly game.

On the supply side it is argued that:

(1) capacity creation and the capital market,
(2) learning by doing and further technological advance, and
(3) the nature of the oligopoly game

are relevant factors that impinge on the diffusion path. Empirical work suggests that the interaction of such forces will tend to generate sigmoid diffusion paths. The conflicting nature of the models, however, indicates that we have not yet been completely successful in integrating all the different approaches to generate a complete understanding of how these paths are generated. One key theme in all the analysis, however, is that the more profitable it is to use a new technology, the earlier it will be used. The problem that arises, however, is that the factors that affect profitability are endogenous to the diffusion process.

Notes and references

1. Davies, S. (1979), *The Diffusion of Process Innovations*, Cambridge, Cambridge University Press. At the time of writing, a survey by David, P., was forthcoming in the *Journal of Economic Literature*.
2. See Davies, op. cit.
3. Mansfield, E. (1968), *Industrial Research and Technological Innovation*, New York, Norton.
4. Glaister, S. (1974), 'Advertising policy and returns to scale in markets where information is passed between individuals', *Economica,* 41: 139-56.
5. Spence, A. M. (1981), 'The learning curve and competition', *Bell Journal of Economics,* 12: 49-70.
6. Metcalfe, J. (1981), 'Impulse and diffusion in the study of technical change', *Futures,* 13: 347-59.
7. Chow, G. (1967), 'Technological change and the demand for computers', *American Economic Review,* 57, 1117-30; Stoneman, P. (1976), *Technological Diffusion and the Computer Revolution*, Cambridge, Cambridge University Press.
8. Spence, A. M. (1979), 'Investment strategy and growth in a new market', *Bell Journal of Economics,* 10(1): 1-19.
9. Davies, op. cit.
10. Stoneman, P. (1981), 'Intra-firm diffusion, Bayesian learning and profitability', *Economic Journal,* 91: 375-88; Stoneman, P. (1980), 'The rate of imitation, learning and profitability', *Economics Letters,* 6: 179-83; Lindner, R., A. Fischer and P. Pardey (1979), 'The time to adoption', *Economics Letters,* 2: 187-90.
11. David, P. (1969), 'A contribution to the theory of diffusion', Stanford Center for Research in Economic Growth, Memorandum No. 71, Part II, p. 10.
12. Defining $1/\omega = C \cdot R/W$, i.e., relative prices of inputs.
13. Davies, op. cit.
14. Stoneman, P. and Ireland, N. (1981), 'The role of supply factors in the diffusion of new technology', paper presented at the Association of University Teachers of Economics Conference, University of Surrey.
15. Spence, op. cit.
16. Reinganum, J. (1981), 'Market structure and the diffusion of new technology', *Bell Journal of Economics,* 12(2): 618-24.
17. Romeo, A. (1977), 'The rate of imitation of a capital-embodied process innovation', *Economica,* 44 (1): 63-70.

10 On the adoption of technological innovations in industry: superficial models and complex decision processes*

Bela Gold

Introduction

The technological competitiveness of firms and industries is determined not by the rate at which significant innovations are developed, but by the extent to which they are applied to commercial operations. The importance of such adoption decisions is further emphasised by the fact that evidence of resistance to the utilisation of demonstrably effective technological advances tends to discourage managerial commitments to risky and costly efforts seeking additional advances.

Widespread recognition of the importance of understanding the factors affecting adoption decisions is apparent from the considerable array of publications dealing with them. Unfortunately, however, most of these publications have provided only very limited, and even misleading, insights into the determinants of actual decisions about specific innovations in real firms. Such inadequacies seem to have been due in large measure to an understandable, but nevertheless crippling, premature emphasis on the formulation of broadly applicable generalisations. This has encouraged reliance on relatively superficial concepts and methodologies as well as on highly vulnerable samples both of statistical data and of managerial judgements.

It would appear useful, therefore, to review the key shortcomings of findings published so far and to add the tentative results of an extensive array of empirical studies conducted by our Research Program in Industrial Economics at Case Western Reserve University. These may help to highlight the urgent, but still unmet, needs of policy-makers in industry and government who are concerned with re-invigorating the technological competitiveness of producers. The related analysis and discussion may also help to meet several other needs, including: the correction of misleading expectations concerning diffusion rates; the displacement of erroneous measures of the 'satisfactoriness' of such rates; the provision of a more systematic framework for managerial evaluations of prospective

---

*Most of this chapter appeared under the same title in *Omega* (1980), 8(5): 505–16, published by Pergamon Press. It is reproduced here by permission of the Chief Editor of *Omega*, Professor S. Eilon.

innovations; and the development of more effective guides for governmental efforts to determine which technological innovations should be encouraged to diffuse more rapidly, or more fully, and also to devise incentives to promote such objectives.

On the decision-making context of innovation evaluations

Most research on innovation-adoption decisions and on diffusion rates has been seriously undermined by over-restrictive as well as unrealistic conceptions of both the external and internal aspects of the decision-making framework that is involved.

The external context: simplifying assumptions v. realities

Most econometric models of the diffusion of technological innovations are based on the erroneous conception that the diffusion process is like filling a bottle. Thus, it is supposed that a specific innovation is progressively adopted by an unchanging and essentially homogeneous population of potential users. Such prospective users are assumed to have fixed and basically similar objectives, operations, products, decision processes and evaluative criteria; they are expected to differ significantly only in respect to their respective estimates of the technological risks and profitability of the innovation, the availability to each of the capital required for adopting it, and the attitudes of their managements towards technological and other innovations in general. Accordingly, diffusion rates are expected to change over time as a result of adjustments in the costs of adopting the innovation and in the availability of capital to non-adopters. In addition, the proportion of adopters is expected to increase with evidence of decreasing technological risks and also with growing competitive pressure from earlier adopters.

But each of the fundamental elements of this conception is unrealistic. First, far from being essentially fixed, almost every technological innovation in industry undergoes numerous significant changes in its service capabilities with time. These may affect such factors as reliability, operating flexibility and efficiency, precision and other aspects of the quality of performance, applicability to specialised purposes, and hazards in use. Moreover, these are usually accompanied by changes in investment requirements and operating costs. Hence, 'the innovation' seems to be a fixed entity only to those who are ignorant of its technology and specific applications. To the engineers who seek to increase diffusion of the original innovation by improving and adapting it to the needs of expanding sectors of prospective users—and there may be hundreds or even thousands of technical specialists among the domestic and foreign firms involved in relevant development efforts—'the innovation' constitutes a system of continuously changing potentials and limitations.

It should also be noted in this connection that the variety of forms and capabilities and economic effects involved in the metamorphosis of any major technological innovation pales in comparison with the staggering diversity encompassed by the term 'technological innovations'. The fact that these may differ with respect to virtually every characteristic which is likely to have economic effects further undermines the persuasiveness of the numerous studies which have sought to derive general diffusion patterns and general models of adoption decisions on the basis of oddly assorted, as well as miniscule, samples of technological innovations.[1]

A second general shortcoming of the traditional approach to the diffusion of technological innovations is the assumption that the population of potential users is readily identifiable and essentially fixed. Actually, most studies have failed even to identify the group of prospective adopters realistically because of an understandable eagerness to utilise the convenient industrial categories offered by available statistical series. But this tends to reflect either a wishful or an ignorant underestimation of the numerous and important differences among the plants included within most of these categories. Among such widely prevailing differences which may affect the potential benefits of particular innovations—or of successive modifications of them—important ones include product designs, product-mixes, the patterns of make-or-buy arrangements, equipment characteristics and modernity, scale of production, quality standards, and various locational advantages and disadvantages involving access to needed inputs and markets. Thus, at any specified time, the current state of development of a particular innovation is unlikely to be directly relevant to the needs of all plants encompassed by the industry categories for which statistical data are readily available.

Moreover, the population of prospective adopters tends to change over time because of changes in the range of sizes in which 'the innovation' becomes available, as well as in its service capabilities and limitations. Combined with concomitant adjustments in its investment requirements and operating costs, such innovational modifications may increase its attractiveness to additional groups of firms within, and also beyond, the assumed relevant statistical category. Indeed, the fundamental conceptual inadequacy of the traditional, essentially static, view is the failure to recognise the powerful interacting pressures for continuing change exerted on the developers of the innovation to extend its range of applicability in order to expand potential markets and on prospective adopters to reappraise the value of such increasing capabilities.

A third basic weakness of many models of technological diffusion, especially those based on the projection of past statistical data, is their pervasive tendency to ignore significant dissimilarities in the economic conditions which characterised the various periods. A monopolising concern with the pattern of changes in the quantitative magnitudes constituting the selected statistical series results in overlooking the accompanying changes in business cycles, inflation levels and even such critical changes in the firms and industries involved as are represented by growth rates, profit levels, labour problems, regulatory pressures and input shortages. No effective methods for eliminating the influence of all such factors on diffusion rates have been developed as yet, despite a variety of expedients and much effort; such techniques would be of doubtful value even if successful, for there is little interest in determining what diffusion rates might have been under improbably fixed economic conditions. However, valuable analytical insights might well result from determining the effects of various of these conditions on diffusion rates. For example, one of our studies suggested that, contrary to our expectations, diffusion rates did not increase consistently during the periods when new capacity was added.[2]

In short, there is ample basis for doubts about the validity and usefulness of the findings of most models of technological diffusion in industry because of their reliance on oversimplified concepts and heterogeneous samples.[3] Hence, the resulting empirical findings should properly be regarded as individual descriptions rather than as broadly applicable analytical generalisations. Moreover, most such findings cannot even be accepted as effective descriptions, except with the explicit understanding that they cover the relationship between a crudely defined cluster of innovations and an even more ambiguous conception of potential adopters.

It would seem to follow, therefore, that the saturation curve so widely used to depict diffusion patterns, or to assess diffusion rates, or to estimate shortcomings in diffusion levels, is misleading. Instead of one fixed estimate of potential users for the entire period of diffusion, with which actual adoption levels are compared, such charts should show the successive increases in the population of potential adopters associated with important changes in the applicability and economic benefits of the original innovation. Thus, Figure 10.1b would seem to be more appropriate than the traditional Figure 10.1a, and Figure 10.1c might be even more accurate in noting that successive stages of innovational developments are associated with changes in potential users as well as actual adopters. The most significant implication of this conceptual

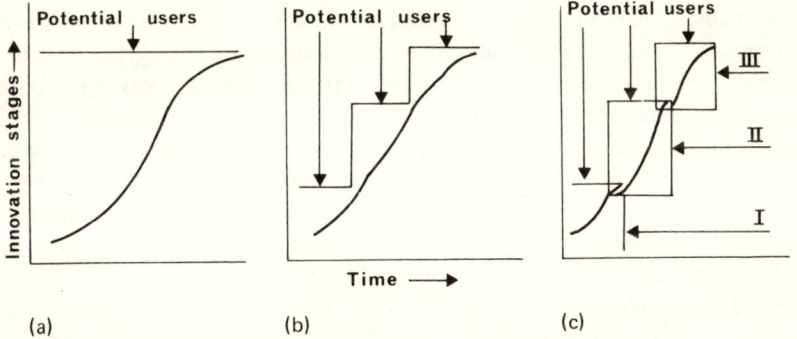

Figure 10.1: Alternative conceptions of innovation diffusion patterns

reorientation is that more effective understanding of diffusion patterns and of the factors affecting them requires more knowledgeable estimates of: the population of prospective users in any specified stage of an innovation's development; the number of additional adoptions likely to result from changes in the input and output pressures on prospective users, even without significant improvements in the innovation; and the number and kinds of additional adopters likely to be attracted by alternative further technological improvements in the innovation, and by reductions in its investment requirements and operating costs. Incidentally, careful research has also led to a serious questioning of the traditional use of sigmoid curves to depict diffusion patterns.[4]

The internal context: simplifying assumptions and realities

The first grossly unrealistic assumption about the intra-firm context of decisions involving the adoption or non-adoption of technological innovations involves the arbitrary foreshortening of needed analytical perspectives. This results from focusing immediately on evaluations of a particular innovation—and how such evaluations might differ among those firms which adopt it and those which do not. But such an approach ignores the frequently dominant, and always important, role of the pre-decision environment. For any period to be covered by current managerial planning and commitments, this covers the specific nature and relative urgency of all of the needs to be dealt with, the availability and relative net advantages of non-technological as well as technological options, and the availability of technical, managerial and financial resources to implement such alternative measures.

Firms within the same general sector of industry may well differ in the relative urgencies of pressures to increase sales, improve product quality, reduce inputs in short supply, lower production costs, and so on. Because of such divergent needs and disparate resources as well as dissimilarities in managerial strategies, there is no basis for assuming that all firms in an industry are seriously considering the adoption of any specified technological innovations within any given period—much less that they are all considering the same innovation.

A second grossly unrealistic simplifying assumption concerning the intra-firm context of decisions about adopting technological innovations is that such decisions are based on expected profitability after adjustments for probable risks and uncertainties. This is merely a tautology rather than a useful analytical insight, for it says in effect that a profit-seeking firm makes decisions which seem to favour profits and, hence, that if an innovation is adopted, it is expected to be profitable. As a matter of fact, however, diffusion patterns suggest that numerous firms arrive at quite different evaluations of the same innovation more or less simultaneously. Hence, effective understanding of diffusion patterns requires more thorough study of how expected 'net benefits' are estimated and to what extent such 'evaluations' are really only rationalisations of decisions arrived at on less obvious and less objective grounds.

Despite frequent casual references in capital budgeting discussions to estimating the 'profitability', and even the 'net present value', of technological innovations and other major capital projects, it is extremely difficult to make such estimates within reasonable margins of error. This is due in large measure, of course, to uncertainties about future changes over five or ten years in the level and product composition of market demand, in product and input factor prices, in competitive pressures, and in other important determinants of profitability.[5] But it is even difficult to make such determinations after the innovations have been adopted and put to use because of many interactions with concomitant changes in other internal operations and management policies as well as in product and factor markets. It is instructive to note therefore that Nabseth and Ray's collection of eight studies of the diffusion of industrial innovations reports that expected profitability was seldom given as a basis for adoption decisions and seldom demonstrated after periods of use.[6]

Similar questions can be raised about the frequent casual references in the literature to estimating risk. But how is it done—and how well? One may readily grant that the risk of technological failure tends to decrease with increasing evidence of successful operations. But how effectively can estimates be made of the possibility of further superior improvements after an innovation has been adopted? Even more important, how well can management staffs estimate the wide array of economic risks that may be confronted, such as those cited earlier? At any rate, field research readily demonstrates that many innovations are adopted not in the hope of increasing profitability, but in order to minimise reductions in profitability threatened by competitors' advances or by prospective special disadvantages facing the given firm.

A third grossly vulnerable basis for studies of decision-making about technological innovations in individual firms is the widespread reliance on *ex post* findings and interpretations. Some well-known studies are even unclear about how long ago the relevant decisions were made (often 10 to 20 years ago or more), and about whether the respondents cited were effectively involved in making such decisions (often not). The results of such studies are accordingly open to serious misinterpretations because of the enormous differences between hindsight perspectives and expectations about the unknown future. For example, hindsight judgements tend to stress *ex post* criteria instead of those which loomed largest when the decisions were made; hindsight evaluations are also more likely to rationalise whatever results were actually realised, crediting favourable outcomes to sound decisions while blaming unfavourable outcomes on external developments. Moreover, the judgements of current executives about long past decisions often reflect much unintentional forgetting as well as unacknowledged ignorance. And even the most serious efforts to evaluate the effects of long past decisions can hardly make effective allowances for the effects of interactions with intervening internal and external developments, many of which are likely to have been unexpected.

Finally, attention should also be drawn to the vulnerability of the various studies which have used heterogeneous samples of firms in an effort to identify those characteristics most frequently associated with responsiveness to innovational opportunities. The most effective evaluation of this general approach is Uhlmann's study of twenty characteristics of 126 firms in Germany, Sweden and the United Kingdom covering their responses in a total of 218 cases of innovation representing eighteen distinctive types of innovation. Although he found that a number of the individual characteristics were associated with responsiveness to innovations on a statistical basis, no firms represented a substantial combination of such characteristics—indeed their actual patterns were virtually unique, both as among firms in the same sectors of industry and also as among those responding to particular groups of innovation.[7]

Actual decision-making processes

Evaluational bases for decisions

One of the major gaps in our understanding of decisions affecting the adoption of technological innovations concerns the bases for the evaluations on which such decisions are based. Few studies even bother to cover the expected net results beyond claiming, or merely implying, that improved profitability was expected. But this reveals nothing about the specific arrays of estimated benefits and burdens which underlie such summary evaluations, nor about the bases for such judgements, nor about the margins of uncertainty considered to be associated with each of these estimates. Nor does it provide needed insights into the technological alternatives which were considered and the bases on which they were rejected as less attractive. Even more important, there is seldom any analysis of the changing evaluations over time as a result of which prior decisions to reject a given type of innovation are eventually reversed. Indeed, one cannot avoid suspecting that some (and perhaps most) of the limited number of ostensible analyses of the decisions leading up to the innovations being reported on are *ex post* rationalisations heavily interwoven with hindsight judgements and omissions.

But efforts to develop sounder diffusion processes—which may increase adoption rates for some innovations while reducing them for others—require fuller awareness of the criteria which are commonly employed, how they are estimated, and how they are weighted in formulating final evaluations. Only by identifying such elements, and then appraising the vulnerability of the means employed to help guide decisions, can attention be drawn to the relative weaknesses of technological and economic estimates of prospective results—as a basis for seeking to improve both.

For example, our field research suggests that the most common sources of errors in estimates of the expected technological benefits of innovations centre on:

(1) underestimating the time needed to achieve effective functioning of innovation, often by a considerable margin;
(2) overestimating the average utilisation rate as a basis for appraising benefits;
(3) underestimating the need to make adaptive adjustments in the preceding and subsequent operations of an integrated production operation—or in the reallocation of orders and support resources between a new facility and older facilities devoted to similar operations; and
(4) underestimating the problems and costs of gaining labour acceptance of associated changes in tasks.

Economic evaluations of the expected effects of technological innovations are frequently based on a wide array of simplifying assumptions, among which the following seem to be the most common and most influential:

(1) that expected reductions in man-hours per unit of output will be accompanied by roughly comparable reductions in unit wage costs;
(2) that expected reductions in material requirements per unit of output will yield parallel reductions in their unit costs; and
(3) that resulting costs savings can be carried over into increased profits.

But the first two assumptions ignore the tendency for changes in unit input requirements to interact with factor prices. Thus, increases in output per man-hour often engender comparable increases in wages per man-hour whether through piece rates or through trade-union demands, thus tending to minimise expected reductions in unit wage costs. Moreover, trade unions often resist the layoffs made possible by technological innovations. Indeed, our research reveals that some companies in the United States no longer permit the inclusion of expected wage costs savings in justifications for proposed capital projects, on the grounds that these all too frequently prove unrealisable.

In the case of reductions in material requirements per unit of output, the effects on their unit costs seem to be more variable. When such reduced inputs are attributable to tighter specifications of quality or dimensions, price increases may offset the expected cost savings. When such reduced inputs have been made possible by an innovation which is being adopted by competitors as well, the expected cost savings may be accentuated as a result of the depressing effect on the price of such materials due to widespread reductions in demand. But cost savings attributable to shifts to lower-priced materials may shrink in time as competitors also turn to such substitutes and thus increase the demand for them along with attendant prices. Estimates of expected savings in unit material and unit wage costs are also often erroneous because of the tendency to base them on current input prices instead of taking account of recent trends in such prices. Thus, even demonstrable reductions in the quantity of unchanged materials per unit of output may serve only to slow the rate of continuing increases in such unit material costs instead of yielding actual reductions.[8]

It is also very important to recognise that the increasing diffusion of cost-saving innovations under competitive conditions tends to generate reductions in product prices as producers struggle to maintain market shares. Thus, the profit margins of early adopters are likely to undergo progressive shrinkage over time. In short, each of the foregoing considerations stresses the importance of seeking to estimate the changing pattern of favourable and unfavourable economic effects period by period over the expected life of the innovation instead of simply multiplying current estimates of annual savings by the expected economic life of the undertaking.

In turn, these perspectives necessitate facing up to the serious inadequacies of the long-range economic forecasts which are at the core of capital budgeting evaluations of innovational and other major project proposals. Attendant margins of error are likely to be quite large even in forecasts for entire industries, to say nothing of the even greater hazards involved in forecasts for individual firms and even single plants.[9] In view of the extreme difficulties involved, one can readily understand the resulting tendency to project current moods of optimism and pessimism into forecasts of ten years and longer,[10] and one may even sympathise in some degree with the widespread practice of projecting the trends of the past 10 to 20 years into the next decade or two, despite the repeated demonstrations of major dissimilarities between the past and the future.[11] But such tolerance should not prevent awareness of the fundamental shortcomings of available forecasting methods and the need, therefore, to avoid placing heavy reliance on them. In particular, it would seem useful to consider minimising dependence on statistical and econometric forecasting in favour of intensive analysis of the specific pressures and opportunities which lie ahead, and to recognise that success is more likely to result from alert recognition of, and effective adjustment to, the inevitable emergence of unexpected developments than on the correctness of the original forecasting.[12]

Decision and revision processes

It is commonly implied that decisions to adopt innovations somehow emerge automatically whenever continuous evaluations of an array of innovational alternatives finally identify candidates which exceed the current 'hurdle rate' for new undertakings. But there is no basis for this conception of a ceaseless economic radar scanning of possibilities of all kinds until one or more return a sufficiently profitable echo. Although many firms turn at least a casual eye towards any purportedly exciting new prospects on the horizon, serious evaluative efforts are much too costly to be applied indiscriminately. Hence, although a scanning awareness of current developments represents a minimum requirement of recognising truly promising targets, most intensive commitments to evaluate particular possibilities tend to be triggered by such circumstances as the following:

(1) a threat to current market share resulting from technological advances by a competitor yielding improvements in product capabilities or prices which are patently attractive to current or prospective customers;
(2) a progressively weakening competitive position which requires consideration of developing or adopting risky and costly new technologies as the most promising remaining means of safeguarding survival;
(3) a recent experience involving a technological innovation which yielded substantial competitive advantages as well as increased profitability and thus engendered greater confidence in the practical potentials of additional such undertakings; and
(4) persuasive evidence of the imminent commercial applicability of an internally developed technological innovation promising important market benefits.

The point to be emphasised is, as was suggested above, even the most thorough *ex ante* estimates of the prospective net benefits of adopting innovations tend to be subject to wide margins of error. Whether resulting decisions are favourable or unfavourable to adoption depends, therefore, on the managerial judgements which are invoked as supplements to these ostensibly objective, but patently vulnerable, estimates. Such judgements seem likely to reflect the value biases derived from the past training, expertise and experiences of various executives.[13] But most executives would nevertheless prefer to defer reliance on such hazardous guides in making major adoption decisions until external pressures threaten rising penalties for continuing delays. As a result, it is not at all uncommon to find that important innovations have been under consideration for several years before being adopted, thereby further emphasising the importance of learning more about the factors which finally trigger adoption decisions, if our understanding of diffusion patterns is to be significantly improved.

Another oversimplification involves treating adoption decisions as though they were climactic once-for-all actions rather than representing only initial commitments subject to successive modifications on the basis of further information and experience. Of course, innovations differ in the rapidity with which investment decisions are carried out and in the extent to which later adjustments can be made in basic processes, in equipment capabilities, and in product characteristics without forbidding costs.[14] But undertakings that take one or two years to complete are likely to permit repeated modifications of original decisions, and this is even more true of projects taking longer periods.

Recognition of the continuing dynamics of these decision processes has two important implications. It emphasises the potential errors of inferring the bases for the original decisions from the eventual results—and also of evaluating performance through comparison with the originally defined objectives. In addition, this broader conception highlights the need to study the extent to which later adoptions by firms with multiple operations susceptible to similar applications are based on changing evaluations of the innovation's capabilities, as well as of market pressures and of internal urgencies.[15]

Appraising innovational effects

How effectively are the results of technological innovations determined? Such findings would obviously tend to have an important bearing on the rate of diffusion inasmuch as evidence of significant rewards is generally regarded as the most powerful incentive to increasing diffusion. Oddly enough, however, there is an astonishing paucity of published research evaluating the actual effects of such innovations after they have been installed, in contrast to the considerable literature on estimating the probable effects of technological innovations before decisions are made to adopt or reject them.

Inquiries suggest that the reason for such neglect is the widespread assumption that the purposes, methodologies, applications and interpretations of such undertakings are so obvious as to offer no interesting problems. Our field research, however, yields the strongly contrasting view that such post-installation evaluations are shot through with difficult problems and dubious bases for many of the results which are reported within firms.

Analysis of the problems and effectiveness of post-installation evaluations of the effects of technological innovations offers several potentially important contributions to the management of innovational processes. To begin with, such appraisals could provide a direct comparison of results with the expectations which led to adoption decisions. Even more important, comprehensive evaluations could explore the specific loci and causes of deviations between expectations and results, thus indicating the relative accuracy of various component estimates and also identifying any variables which were ignored. Moreover, such evaluative efforts may reveal insensitivities in the performance measurement system to innovational impacts, thus counselling changes in order to minimise consistent anti-innovation biases.

Some shortcomings of current evaluation efforts

Only a limited array of post-installation evaluation methods has emerged as a by-product of our field research on the effects of technological innovations. Hence, the following judgements must be regarded as preliminary and tentative indications of possible shortcomings. Most of the 'make good', 'follow-up' and 'post-audit' evaluations examined concentrated primarily on simply measuring actual results, including: the costs of acquisition, construction and installation relative to budget; the acceptability of technical performance; and resulting operating costs. Except for comparisons with allowed budgets and expected total unit costs, few methods were characterised by comprehensive comparisons of actual input requirements, factor prices, product quality and price, output levels and other aspects of performance with the respective estimates which led to the decision. Especially glaring is the common failure to consider the time required to achieve the 'acceptable' levels of performance before evaluation efforts tend to be initiated relative to expectations. Also disappointing is the virtual absence of any systematic efforts to 'learn from results' as a means of improving *ex ante* estimates of the effects of prospective technological innovations in the future.

Our explorations suggest, in addition, that formal evaluation efforts, except for comparisons of actual expenditures with allowed budgets, seem to be much less common with respect to very large projects—especially when their results seem unfavourable. In explaining such lapses, the two most common reasons given were that each such project is necessarily unique and hence evaluations would have no feedback value in considering future projects, and that there was no interest in 'spilled milk' or in 'flogging dead horses'.

One of the common limitations of post-installation evaluative efforts has been reliance on an over-restricted framework of considerations. A major part of the decision-making process involves choosing among a variety of available technological and non-technological means of meeting the most urgent needs of the firm at the time of decision. But a comparison of actual results with expectations with respect to the final choice made throws no light on the accuracy of the evaluations which led to the discarding of the other alternatives considered. Nor do such comparisons reveal the effects of having adopted innovations which succeeded in easing what were considered urgent needs at the time of decision at the cost of neglecting other needs which proved more serious. In short, there has been an almost complete failure to evaluate the substructure of evaluations that then determined which of the alternative means of dealing with specified needs should be adopted.

In the case of very large projects, the tendency to seek out and to emphasise favourable aspects of results seems to be attributable to a concern that negative judgements would reflect on high-level officials and be resented by them. Such biases are often built into the evaluation process because allocation of such responsibilities to the officials deemed to have the relevant expertise often involves reliance on those who were also involved in project proposals and decisions. Thus, technical evaluations are usually left to engineers and various cost estimates to the respective specialists—partly because of the absence of effective internal alternatives and partly to protect the confidentiality of findings. Moreover, those assigned to making such evaluations are often led to mute critical judgements lest these inhibit future co-operative relationships with the officials responsible for the project. Indeed, we have not yet encountered any cases of wholly independent evaluations involving technological and economic competence. The seriousness of this problem is indicated by the fact that a senior officer of one of the major steel companies told us that they have abandoned such post-installation evaluations because they were invariably found to be so biased as to render them of dubious value.

A third set of shortcomings of post-installation evaluations arises from the time focus and criteria employed. For example, most cases examined relied on a single narrowly focused evaluation made within 6 to 12 months of the project's completion. These early estimates tended to yield over-optimistic findings because generous allowances were usually made to offset shortcomings which were assumed to be attributable to temporary difficulties in achieving effective operations, to temporarily increased maintenance problems, and to temporary under-utilisation. Such early evaluations also tended to be inadequate because it is only after the innovation has achieved effective functioning and reasonably high levels of utilisation that efforts to maximise realisation of its potential lead to adaptive adjustments in preceding and subsequent operations, and even to possible modifications in product designs and product-mix. Hence, more effective appraisals would require successive evaluations every six months for at least three years (or even longer if effective functioning has not yet been achieved—as in some cases of continuous casting, for example) to ensure effective determination of practically sustainable performance levels, and to ensure coverage of the wider repercussions of the innovation.

In addition, the actual effects of technological innovations are often measured inadequately because cost accounting categories are not revised to reflect important aspects of the innovation's contributions. These may include changes in the quality of the inputs used, in the nature of the processing or fabrication performed, or in the service capabilities of products. Other significant effects which are commonly disregarded include changes in the flexibility of operations which can be performed and changes in the precision with which processes can be controlled. Moreover, concentration on the comparison of results with expectations tends to result in inadequate probing of the specific causes of shortfalls. As a result, observed deficiencies are all too readily ascribed to unpredictable or external factors. Such minimising of internal shortcomings obviously prevents identification of targets needed for improvement efforts. Finally, attention should be directed to the seemingly universal avoidance of estimates of the innovation's incremental contribution to profitability. This implied recognition of the difficulties in attempting such evaluations, even on the basis of actual *ex post* data, raises even more serious doubts about the usefulness of *ex ante* estimates of such profitability effects as a basis for adoption decisions.

Installation evaluations

Efforts to improve evaluation of the post-installation effects of technological innovations are confronted by a variety of questions concerned with how to reduce biases engendered by common production, costing and other practices. For example, means need to be considered for reducing the favourable biases associated with at least five common production practices encountered in our research. One of these involves shifting the most advantageous orders and the best operators to the new facilities, along with granting them top priority in access to ancillary facilities and to repair and maintenance services. Another such practice involves maximising the utilisation rate of the new facilities at the expense of the older facilities. Such biases also result from motivating greater labour efforts and care through providing improved pay incentives and working conditions. Still another source of differential advantage frequently involves improving the quality of work inputs from preceding operations and increasing the standardisation of the tasks to be performed by the new facilities. And the question must also be faced of how long additional improvements to the initial installation are to be attributed to the original innovation instead of to subsequent innovations.

Parallel problems are also confronted in seeking to minimise unfavourable biases in evaluating the effects of innovational decisions attributable to production conditions. One of these involves underutilisation due to a recession. Another involves the effects of unexpected deficiencies in the quality of the materials to be processed. And a third may be caused by the necessity of modifying product specifications to adjust to changing customer preferences.

An important group of problems relating to proper evaluation of the cost effects of technological innovations concerns whether the following should be treated as increases in the investment embodied in the innovation or as current additions to operating costs:

(a) additional outlays in order to improve the effectiveness with which the new facility functions;
(b) the cost of interruptions to production caused by introduction of the innovation;
(c) the cost of delays before achieving effective functioning of the innovation, including the cost of modifications, 'debugging', training operators and trial runs; and
(d) the costs and outlays involved in readjusting preceding and subsequent operations in order to achieve effective integration with the capabilities of the innovation.

A related problem concerns whether to credit the innovation with cost reductions only in its own operations, or also to credit it with all cost benefits resulting from the adaptive improvements made in other operations, including procurement and engineering. Still another problem concerns how to evaluate the contributions of the innovation to changes in revenues apparently associated with innovation-induced adjustments in product quality and in the flexibility of product-mix.

Perhaps the most difficult problems of all involve seeking to disentangle the effects of the innovation from those of a wide array of concomitant developments. Among these, internal developments might include the introduction of other technological and non-technological innovations, as well as changes in management policies relating to prices, marketing, labour relations and other factors affecting competitive position. External developments might include changes in industry supply-demand relationships, changes in the availability and prices of input factors, technological and other innovations by competitors, and modifications in government regulations affecting the industry.

Finally, because many technological innovations require investments which are likely to be embodied in them for 10 to 20 years or longer, some attention must be given to the problem of longer term evaluations. Because relevant costs, revenues and net investment tend to change from year to year, evaluations of an innovation's effects would also yield changing results over time, quite possibly involving substantial changes in their favourableness. Does this mean that all project evaluations should be repeated annually? For how long can their effects be differentiated from the combined impacts of all other developments? Three other questions seem to be even more fundamental:

(a) What would be the significance for current decision-making of learning that some past decisions yielded favourable results in the short run, but unfavourable results after five years—whereas others yielded the reverse pattern of results?
(b) What margins of error are likely to be associated with the 10 to 20 year estimates of output, costs, prices, interest rates and profits used in capital budgeting models—and should estimates subject to wide margins of error be used as the basis for decisions to adopt or to reject technological innovations?
(c) And if the preceding question is answered in the negative, what alternatives are available to management?

Strengthening ex post *evaluations to help improve future decisions*

The actual results of post-installation evaluations of technological innovations are likely to have little effect on current decisions concerning the adoption of new innovations. This absence of a feedback effect is likely to be true in part because *ex post* evaluations tend to vary significantly during the early years after installation, and because there is considerable awareness of the biases commonly reflected by initial evaluations. But the influence of later evaluations also tends to be minimised for two other reasons. Eventual determinations of the actual effects of long past innovational decisions are regarded as increasingly irrelevant to the different innovations and altered urgencies faced in later years. Even more important and instructive, however, is the tendency to view the majority of such eventual results as being attributable in larger measure to the effectiveness of management policies during the years in which the innovation was utilised than to the carry-over effects of the original decision. In short, the limited usefulness of the kinds of *ex post* evaluations which have been encountered helps to explain the essentially peripheral interest of many managements in such exercises.

Such past shortcomings, however, have prevented realisation of the valuable potentials of revised approaches to post-installation evaluations. Necessary revisions should include comparisons of the actual results with estimates: the probable results of having rejected the innovation, or of having delayed its adoption by one, two or three years; comparisons of actual results with the expected results at the time of the original decision; and comparisons with the apparent results of the technological and other innovational decisions made by competitors at the time when this firm made its original adoption decision. Revised approaches should then seek to explore the causes of the differences revealed by the preceding comparisons. In particular, it would be instructive to identify which differences were attributable to technological, economic or market factors; which of these represented internal rather than external developments; and finally, which might reasonably have been predicted at the time of the decision and which were clearly unpredictable.

As a result of these more comprehensive insights into the complex patterns and multiple determinants of the actual effects of technological innovations, consideration might be given to enriching the past objectives, coverage and methods of *ex ante* evaluations. For example, in defining the objectives on the basis of which choices are to be made among alternative innovations, the need might well be recognised to dig beneath the generalised objective of improving profitability and to concentrate more sharply on the specific product, process, cost and other adjustment targets involved in bettering past performance—thereby providing more precise criteria both for choosing among the options being considered, and also for evaluating post-installation results. Moreover, instead of merely comparing the relative net benefits of alternative innovations, efforts should be made to clarify the technological and economic assumptions underlying them, and also to indicate the margins of error likely to be involved—including any relevant references to the results of *ex post* evaluations. More particularly, evaluations of prospective technological innovations should seek to specify: the sources and expected magnitudes of the estimated superiority of recommended innovations over current facilities (whether introduced as replacements within present plants, or, if relevant, as parts of new plants); the factor price and other assumptions involved in converting expected technological improvements into economic benefits; and estimated advantages and disadvantages of adoption now versus deferring adoption, including specification of associated assumptions concerning the concurrent behaviour of competitors.

Another by-product of attempts to determine the *ex post* effects of technological innovations more effectively may be recognition of the need to change some of the categories commonly used to assess the productivity and cost effects of prospective innovations. Specifically, measures must be designed to take account of changes in input qualities, the nature of processing requirements, the shifting of processing tasks to other operating units, improvements in process flexibility and alterations in product quality. These tend to alter both the physical magnitudes and economic value of productive contributions and yet have been largely or wholly ignored by prevailing measures, which focus solely on changes in input and output quantities, assuming no significant changes in their qualitative attributes.

Concluding observations

The preceding discussion has been presented in order to emphasise the wide range of possible effects to be considered and the variety of viewpoints from which evaluations may be made of the desirability of broader or faster diffusion of any technological innovation in industry. It seems reasonable to argue that industrial managements, government agencies and academic scholars would all benefit from the development of progressively more effective bases for systematic examination and appraisal of the effects of technological innovations as well as of different patterns, rates and levels of diffusion. Results to date have not been of much practical value to industrial managements, partly because decision-making responsibilities tend to limit their interest to findings more directly relevant to their own immediate needs, but in even greater measure because industrial specialists are usually far better informed than transient researchers from outside the industry about the diffusion patterns of relevant technological innovations and the factors influencing them. Nor have the results of research in this area offered much useful guidance to government policy-making because of the recognised vulnerability of attempts to generalise from findings commonly based on patently inadequate samples and obviously inadequate penetration of the complex considerations underlying the actual decisions reflected by diffusion rates.

Perhaps more disturbing from the standpoint of responsible scholarship has been the readiness of some researchers to criticise industrial managements for delays in adopting seemingly relevant technological innovations. Most of these attacks have been based on arguable inferences from data providing hindsight perspectives or comparisons with experience in other countries. But such judgements reflect ignorance of, or lack of interest in, the specialised limitations of the given state of the innovation relative to the dominant needs of prospective adopters in earlier periods. Surely, serious analysts cannot assume away the difficulties of scaling up innovations or adapting them to wider areas of application; nor should they deny the possibility that shifts in factor prices can appropriately alter prior evaluations of innovational effects, or that earlier commitments of available resources to needs deemed more urgent represent rational reasons for deferring adoption of an innovation.

Indeed, it is of fundamental importance to recognise that virtually all innovations involve technological and economic risks at all stages of diffusion, and that decisions to reject may accordingly be entirely justifiable at any stage by some prospective adopters. The fact that most innovations fail to achieve widespread diffusion testifies to the pervasiveness of such risks and the need for evaluating them carefully. It is surely irresponsible in considerable measure, therefore, to launch such criticisms from the safety of hindsight perspectives. Oddly enough, even such *ex post* attacks are seldom buttressed with serious evidence of the economic superiority of the innovations at earlier periods, reliance being placed rather on estimates based on highly oversimplified assumptions which clearly disagree with contemporary evaluations of industry specialists. Accordingly, it would be more in accord with scholarly objectives to try to understand why decisions were made which do not conform to the analyst's expectations than to imply that such decisions were attributable to the sloth, stupidity or ignorance of substantial sectors of industrial management.

Notes and references

1. For a fuller discussion, see Gold, B. (1978), 'Some shortcomings of research on the diffusion of industrial technology' in Radnor, M., I. Feller and E. Rogers (eds), *The Diffusion of Innovations: an Assessment Report to the National Science Foundation*, Northwestern University, Evanston, Illinois.
2. See Gold, B., G. Rosegger and W. Peirce (1970), 'Diffusion of major technological innovations in US iron and steel manufacturing', *Journal of Industrial Economics*, 18 (3): 218-24.
3. For example, see the compendium by Martino, J. (1979), *Development of Predictive Models of the Diffusion of Innovations in Industry*, report to the National Science Foundation, University of Dayton, Ohio.
4. For example, Ray reported in 1969 that in an international study of the diffusion of various technological innovations in industry: 'Neither the curves for individual processes nor their aggregation provided any strong contradictions of this assumption' (i.e. 'that the diffusion curves are linear'). Ray, G. (1969), 'The diffusion of new technology', *National Institute Economic Review*, 48: 40-83. Nor was any support for the general applicability of sigmoid diffusion curves provided by a 1970 publication, covering the first 15 years after commercialisation of the diffusion of 14 major innovations in steel, coal and iron-mining in the United States. See Gold, Rosegger and Peirce, op. cit.
5. For a detailed discussion, see Gold, B. (1977), 'On the shaky foundations of capital budgeting', *California Management Review*, 19 (2): 51-60.
6. Nabseth, L. and G. Ray (eds) (1974), *The Diffusion of New Industrial Processes: an International Study*, Cambridge, Cambridge University Press.
7. See Uhlmann, L. (1979), 'The innovation process: empirical results' in Ray, G. and L. Uhlmann, *The Innovation Process in the Energy Industries*, Cambridge, Cambridge University Press.
8. For illustrations and further discussion, see Gold, B. (1971), *Explorations in Managerial Economics: Productivity, Costs, Technology and Growth*, London, Macmillan, pp. 193-7.
9. For further discussion and some empirical findings, see Gold, op. cit., 1977.
10. See Gold, B. (1964), 'Industry growth patterns: theory and empirical findings', *Journal of Industrial Economics*, 13 (1): 53-73.
11. See Gold, B. (1974), 'From backcasting towards forecasting', *Omega*, 2 (2): 209-23.
12. See Gold, op. cit., 1977.
13. See Gold, B. (1969), 'The decision framework for major technological innovations' in Baier, K. and N. Rescher (eds), *Values and the Future*, New York, Free Press.
14. Such possibilities are considered in Eilon, S., B. Gold and R. Tilley (1973), 'Measuring the quality of economic forecasts', *Omega*, 1 (2): 217-27.
15. See Gold, Rosegger and Peirce, op. cit.

11 Technical advance and trade advantage
D. K. Stout

Introduction

It is now common ground that the static theory of international trade advantage has little to tell us about who will make what in the future. The assumptions that underlie the Heckscher–Ohlin theorem do not typically hold in manufacturing markets. Oligopoly, product differentiation, price discretion and the rapid growth of intra-industry trade are the rule. Multinational enterprises have both increased the mobility of key production factors and, by vertical integration at different locations, separated the principles governing the country of origin of final products from those determining the origin of stages of value added.

Economies of scale have always made a difficulty for comparative advantage theory. Once technical progress is given proper pride of place in the explanation of trade performance (as in the closely related explanation of different rates of economic growth), then economies of growth, position on the learning curve, the growth rates of product markets and the good or bad timing of innovations can tell us more about the global distribution of manufactures and tradeable services than can indigenous factor scarcities.

The international competitiveness of local producers affects both trade flows and the successful unprotected production of import substitutes. New products with unpredictable growth are constantly emerging and displacing old products in the 'perennial gale of creative destruction'[1] upon which profits depend. The population of new products and processes stemming from even one area of technology is myriad, and forecasting exactly which market demands will be satisfied from within which national boundaries is impossible. As in the study of economic growth, and as in the relation of market structure to performance, so here (in a field which draws upon both) one needs to establish principles which apply in persistent disequilibrium. The circumstances are that technical progress permits radical product improvements and drastic changes in factor productivities over periods that are short in relation to the pre-war timetable of international specialisation. In many industries, these changes swamp the effects of relative factor endowments upon global market share. Relative price changes likewise pick up only a part of changing competitive advantage as products are redesigned and transformed.

The characteristic process of developing and securing an industrial power-base is still not very different from what Schumpeter euphorically described in *The Theory of Economic Development*[2] and *Capitalism, Socialism and Democracy*.[3] Typically, advances in knowledge permit an innovation to be made usually by a producer located in an advanced economy and backed by patent protection and a large, sophisticated potential home market. Entry will commonly be delayed, particularly by economies of scale in production and selling, and by the economies of growth experienced along the learning curve. But the lure of the innovator's profit and sales growth, and the awareness of the technology gap that has opened up, strongly motivate entry and imitation. Where production will concentrate, geographically, will depend upon the conditions which prevail in different national economies, and upon market structure characteristics which influence the pace of future technical change. The recording of such influences lacks the neatness of the factor price equalisation theorem, but avoids the empty circularity of 'revealed' comparative advantage.

The key is technical advance, which changes the product population, links hitherto distinct product markets, augments labour and capital in combination and to different degrees, and is, in its effective speed and width of national application, both determined and determining. The recognition of endogenous technical progress lies behind the surge of interest in industrial policy. Let one example suffice: advances in continuous fermentation technology have enabled ICI to manufacture a single cell protein animal feed with ethanol as the raw material and bacteria as the food processors. The effect is to substitute one 600 tonne pressure vessel for 50,000 hectares of soya bean land. A theory of trade advantage founded on national differences in endowments of immobile land, labour and capital and slowly changing production functions cannot encompass these now commonplace events.

Some general conclusions which deserve comment are beginning to emerge from studies of changing competitive advantage. They are these:

— Changing technology is fundamental to changing international location of production, just as it is to understanding the changing futures of competing oligopolists.
— Both innovative activity and speed of diffusion vary with characteristics of the national economies and of the markets within which the innovations occur, feeding back to macroeconomic behaviour and to market structure. These variations have been investigated and described, though no one has so far had much success in modelling and measuring the effects and feedbacks. Thus, the topic remains disorderly, and policy accordingly confused and misguided.
— Trade advantages are enjoyed in those economies where diffusion can be fastest, not necessarily those where the technical leaders reside.
— The market structure that favours innovation and diffusion is only partly

a necessary result of that advantage and the stage of the product's life cycle: it is also capable of being adapted to enhance the rate of technical progress or to hold it back.
— The role of a swarm of small, usually highly specialised 'niche' entrant firms in the embryonic stages of a new high technology is vital. Such firms appear most readily in advanced industrial cultures and help to account for the dominance of those countries in innovation.
— The international diffusion of innovations has more to do with the corporate strategies of powerful competing oligopolists (mainly multinational) than it has with local supply side considerations.
— The social benefits of R & D and innovations whose applications increase efficiency and versatility in potential user industries far exceed the private benefits. There is, therefore, a public policy issue. So far the intervention of governments has in the main been uncoordinated, inconsistent, and often unproductive.
— A new wave of technology redefines both factors and product markets and increases substitutability both from the demand side and the supply side. Conditions of entry and entrenched competitive advantages are thereby changed, with long-run consequences for international specialisation and relative national industrial growth. (Biotechnology now provides a rich source of examples.)

Technology and the theory of comparative advantage

The introduction of technical progress and its rate of diffusion has the same sort of effect upon the pure theory of international trade ('who should make what') as it has upon growth theory. There are two main difficulties which we face in modelling and accounting for differences in the rate of economic growth. First, we cannot assign an independent role to technical progress without actually assuming the answer by arbitrarily specifying the form of the aggregate production function: Hicks-neutral, or Harrod-neutral, or embodied in investment in the Kaldor manner and requiring an increase in the capital–labour ratio to elicit it. Choices between specifying moves along a temporal production function in which technical change is an input or shifts in the function due to exogenous technical progress lead to corresponding arbitrary distinctions between scale economies, effects of capital formation and the independent influence of technical progress. Secondly, 'equilibrium' growth theory, in which the rate of technical progress is taken to be given, is uninteresting. It does not help to explain the differing performance of differing economies.

For this purpose, there are two operationally useful facets of technical progress. In each of these, technical progress depends upon national economic conditions, institutions, market structure and attitudes to change. The first is the speed of innovation: the rate of advance in the state of the art, the rate of product obsolescence and the rate of creation of new products. The second is the rate of diffusion of innovations, individually and severally, through competition within and between national economies and through the planning mechanisms of multinational companies.

It is not only differences in growth rates which the development and spread of new technologies help to explain but, at the same time, relative competitive strength, product by product and industry by industry. Competitive innovative edge is a temporary advantage, blunted by Schumpeterian competitive emulation and resharpened as continuously by successful companies, who thereby sustain or transfer national production advantages (whether by trade or by import substitution). The same disequilibrium processes in which the relative winners are those with the ability to adapt quickly help to explain growth by technical leadership, catch-up growth by jumping technology gaps, and changes in international specialisation.

Innovation and the speed of transfer and diffusion of technology can alter the perceived distribution of trade advantage very quickly. The process is complicated and contains elements of luck in scale and timing. A Concorde may be developed in the wrong decade, a mineral oil-based single cell protein may go into full production after a major upswing in the oil price. Or a billion dollar industry may perhaps develop on the doorstep of a company that has the good luck to be at the leading-edge of recombinant DNA technology at exactly the time of a major discovery about the aetiology of cancer.

Among the manufacturing industries where the pace of technical change is quick is a very large proportion of product markets where the rate of growth of demand is exponential. The phenomenon is familiar. Microelectronics technology, for example, has made it possible to introduce cheap portable new products incorporating microprocessors that satisfy a hitherto unmet or inaccessible need. By contrast, the markets for newspapers or ice-cream or canteens of cutlery grow only gently because these are mainly replacement demands. Technological advance changes the market share of the innovating producer or producing nation by cost reduction or product improvement, resisted by the strong incentives of other established producers to match such improvements in price-performance, so diffusing the changes and limiting the consequent specialisation. Intra-industry trade tends to expand by this process. Completely new products, on the other hand, offer opportunities for very rapid growth of sales to the innovator at this early stage in the product's life cycle and often permit him to achieve scale economies which discourage entry by later would-be followers. In these cases international specialisation can grow rapidly and competitive advantage become entrenched.

There is nothing to be gained in explanation by trying to reconcile these changes *ex post* with the theory of comparative advantage in terms of factor endowments. 'Revealed' comparative advantage is not an explanation, but rather a restatement of the results of successful innovation. No tidy operational model exists to explain or predict either intra-industry trade flows or inter-industry specialisation in conditions of rapid and induced technological change. The Hufbauer model of goods moving through a per capita income hierarchy of countries is much too tidy. Where domestic demand pull is important in inducing R & D effort nationally (for example, because transport costs are high), one can expect to find production develop first in high income countries with a high technology industrial infrastructure.[4] But this tendency does not nowadays have the force of an empirical law. Multinational companies transfer technology and appropriate management and marketing inputs very rapidly into newly industrialising countries, exploiting the potential for upgrading host country skills while wages lag. Even when real wages adjust upwards rapidly, as they have in South Korea and Hong Kong, productivity growth flowing from mass production with the new techniques normally swamps the relative factor price change and the offshore production base stays secure.

It is much plainer sailing to describe the national economic conditions which will make it likely that a particular country's industrial base will somehow expand on the strength of new technology than it is to predict which specific products will succeed. The success or failure of particular products (and hence of developing economies which have, as a public policy decision, put their eggs in one basket) depends upon many loosely related factors. Success or failure follows on from information and development leads and lags, the presence or absence of information-sharing among potential producers, the political popularity rating of particular branches of industry (often leading to overproduction), the complementarity between a new product and other new products, on means of delivery, needs created by environmental changes, the accident of an unexpected change in the cost of close substitutes, the home base of particular multinational companies, and so on. None of this is amenable to the orderly classifications and propositions about trade in an n-factor, n-commodity free trade model under perfect competition. There is no theoretical basis laid for predicting or picking 'winner industries' *a priori*, and that brief fashion in industrial policy-making has now faded.

For an advanced industrial country, historically successful in relatively mature markets where world demand has slowed down, trade performance and industrial growth depend upon creating a climate which is hospitable to innovation, and not upon some logical process of ranking and selection, on the basis of past strength, of a short list of national champions. A country in which everybody plays chess will be more likely to produce a world champion than one which gives hothouse treatment to the best of its few players. Widespread facilities for sport may throw up a pole vaulter or a tennis player. If only tennis courts are provided, the athlete may blush unseen. Growth in a free trade context depends upon the ease with which resources can move quickly to what are turning out to be strengths and leads. The favourable climatic conditions are mobile capital, low risk averseness and especially flexible and versatile human resources, sharply aware of the diverse possibilities for the new products and processes of generic technologies like microelectronics and the fast-developing applications of molecular biology.

Specialisation and stages of economic growth

In newly industrialising countries, education and the capacity for social and institutional change—again the conditions that enhance the ability to change the balance of economic activity and distribute the rewards of higher productivity so that the process of change is welcome—appear to be the secrets of success. It is almost a matter of accident whether a Taiwan, a Korea or a Brazil becomes a major producer of TV tubes or video recorders or digital watches. What is no accident is that some would-be industrialisers are left with cotton textiles, while others can jump into the mass production of goods that hardly existed a decade ago. It is climate, wit and nimbleness that matter and not specialised factors of production, for these can be produced or shipped into place.

The original innovative stage in the spread of new technology always takes place in the most advanced economies (usually still the United States). There are two main reasons for this. First, expected profit depends upon that quick growth of domestic demand for the new product which cannot so easily occur where consumption growth mainly means more people consuming more of a product which has already been available for many years. Secondly, in an advanced economy with a highly developed complementary substructure of manufacturing and service activities, the new product, resting in that ambient technology, is rapidly improved and cheapened. Many unpredicted uses are found for it, in combination with other new products, altogether transforming production possibilities and displacing existing processes.

Generic technologies applied to established areas of demand lead to the development of classes of new product which themselves break down and invalidate traditional market and industry categories. The producer of lasers can either supply, or enter, fields as far removed from the point of view of demand substitutability as visual entertainment, defence, medical supplies, metalworking and supermarket check-out systems. The same is true—emphatically—of very large scale integration in microelectronics. And the same is increasingly becoming true of the genetic manipulation of bacteria and their use in fermentation. Setting aside health care, mineral recovery, sewage and pollution control, and many other industries which genetic engineering will transform, and looking only within food manufacture, the precise tailoring and then mass production of micro-organisms can directly provide alternative feeds, can produce enzymes (working as well outside the cell as inside it) and additives which transform the economics of food-processing, can change the yield or disease-resistance of crops, can control plant disease and infestation without environmental damage and can increase dramatically the efficiency of established chemical routes (like trans-esterification) for extracting nutrients from raw materials. The use of enzymes, biological routes to food flavouring and additives (including proteins), the inhibition of micro-organisms and massive advance in fermentation technology all combine to reduce the minimum production scale of individual food products, to increase shelf-life and to reduce the costs of product differentiation. In the long run, the effect is likely to increase very greatly intra-industry trade specialisation within processed foods among advanced countries.

It is now well attested that industrial growth and export success can develop very quickly in less advanced economies, not just through step-by-step succession to labour-intensive industries like clothing and footwear, but through the application of technology of recent vintage, developed in the west and transferred along with complementary management, training and capital.

As this happens, so-called 'stages of development' in which older, less labour-augmenting techniques are applied in poorer industrialising countries, while the richest economies, like Phoebus, ride the chariot of sunrise, still have some meaning. But the stages can be truncated and some altogether skipped with the help of direct overseas investment by companies near the technical horizon. Stanislaw Gomulka has shown that the creation of a significant technology gap itself constitutes a market signal within the lagging economy, and that, other things being equal, the larger the gap the more intense the competitive pressures to close it.[5] Of course, if that was all there was to the transmission process, then not only would a large part of the difference between different rates of industrial growth and trade performance correspond to differences in the speed with which different economies close in on the moving technological frontier, but these latter differences in speed would depend upon backwardness, and all countries would eventually come together at the moving horizon where catch-up growth would have disappeared for ever.

Looking at the 19 main OECD economies, Robin Marris has recently suggested that three-quarters of the slowdown between 1973 and 1979 in their collective growth rate (compared with 1965-73) could be explained by the 35 per cent increase in richness of these countries relative to the US by 1973 and the consequent reduction in the scope and intensity of market pressure and opportunity for catch-up.[6] The main objections to this test of the process of technology transfer are that the model was so specified that one percentage point of the slowdown in US growth was also ascribed to the 'catch-up' effect, and that if so much differential growth depended simply on backlog, then national growth rates should long since have converged.

The missing factors in the explanation of persisting differences in the ability to exploit abnormal growth opportunities created by leadership elsewhere lie in national differences in institutions and in economic organisation, motivation, government function and social attitudes, all of which affect the speed with which the local skills can change and resources move into fresh fields. Many economists concerned with slow growth and de-industrialisation in Britain, with why others, like Japan, grow much faster, and with the objects and possibilities of industrial policy have come to believe that relative industrial growth depends principally upon the capacity of an economy to displace old products and techniques quickly—upon 'light-footedness'.[7] This is not itself an explanation, but it pushes the hunt for the explanations of emerging industrial and trading strength away both from natural equilibrium growth theory and from static 'snapshots' of specific factor endowments. It diverts it towards the reasons for these national differences in the ability to adapt.

If irremovable obstacles to the speed of change exist in some economies but not in others and if this holds true in the wake of major overlapping waves of technological opportunity such as those that are presently transforming industries internationally—destroying old markets and creating new ones—then economies will tend gradually to converge upon a similar rate of growth (rather than level of productivity). This will happen as the most adaptable catch up with the leader, until eventually restrained by the speed of advance of technical knowledge, while the others move also at that pace, but impeded by obstacles to change within their own industrial structure while they still have unexploited scope to catch up. These more sclerotic economies will be left with small spheres of the new industries and with lower levels of value added per head even than some newly industrialised countries that started from scratch a few decades earlier. In them the optimists are the ones who believe that something can be done about these impediments to adaptation, further multiplied as they are in a vicious circle of falling competitiveness, unprofitability, lack of confidence in future market share, retrenchment, risk-avoidance, and failures to re-equip and embody new technology in new vintages of process and engineering plant and equipment.

Innovative activity and the transfer of technology: some congenial national economic characteristics

It is well known that, among the advanced countries, there is no relation between inventiveness and successful production and marketing of innovations. In the biological sciences, for example, British authors are responsible for seminal research papers out of all proportion to patent activity, venture capital companies started or new products developed. It is impossible to find a unique good proxy for national contribution to technical development. Reported R & D spending often understates the relevant activity; number of patents can be very misleading, particularly where families of closely related patents are applied for, many of which incorporate trivial differences, and where many are put in cold storage; the number of 'important' inventions cannot be known until the history of the industry has unwound. Personnel counts, like number of scientists employed, are usually less significant than how the scientists are combined, how mobile they are, and how they influence company decisions.

International trade advantage in the applications of high technology does not follow from strength in fundamental research. Strong engineering faculties which have close links with industries, large university teams involved in some of the more pedestrian kinds of research like the screening and testing of large numbers of recombinant DNA products, exchange of staff between science faculties and industry, and industrial parks on campus are all helpful but by no means sufficient to ensure such leadership. Yet technical progress at the national level—both in respect of the commercial proving of leading-edge technology and in the sense of national receptiveness to new products and processes developed in North America, northern Europe or Japan—does vary widely between economies, and a lot of work has recently gone into trying to account for these national differences, partly in the interests of improving the progress of mature laggards like Britain.[8]

The two most potent economic factors affecting the rate of innovation are pressure of competition and the expected growth of demand. Both of these influences apply globally, so long as free trade conditions roughly apply, since the markets affected by major innovations are worldwide. Therefore they do not help much to identify the technological leaders beyond the obvious fact that those industrial nations which are strongly export-oriented have an advantage. In the 1980s the two factors are opposed, pulling in different directions. On balance, the increased pressure of competition against a background of slower world growth is likely to speed up the rate of technical progress achieved, because the combat for market leadership is increasingly between powerful champions from different nations between whom the rules of the oligopoly game do not easily permit a cosy agreement limiting the pace of induced obsolescence.[9]

As might be expected where strong feedbacks are at work from innovation to growth and profitability and back to R & D effort, one cannot be categorical about what is a national symptom and what a cause—either of innovative leadership or of speed of diffusion.[10] As Arnold Heertje demonstrates in a classic study of technical change, discriminants (like the rate of investment) which feature as causes of the pace of change in macroeconomic growth models which treat technical progress as exogenous and capital-embodied appear to be symptoms when technical progress is (properly) regarded as endogenous.[11] When invention is treated as a production process, the question arises as to how a particular economy comes to adopt a type and a rate of technical change as distinct from a point on a production possibility frontier which moves outward autonomously.[12]

The ambiguity between causes and symptoms cannot be removed, but national factors more fundamental perhaps than the level of R & D expenditure or the level of patent activity recur in the empirical literature both on innovation and on diffusion. These are:

— the richness and sophistication of the home market;
— the quality, mobility and accessibility to one another of trained scientists and engineers; and
— the congeniality and degree of development of the industrial infrastructure which permits different technologies to converge and combine.

And influencing the speed of diffusion of new products and techniques to 'follower' economies is:

— a political, social and cultural environment which encourages both risk-taking by innovators and the acceptance of change by organised labour.

There is no space to develop these points here and most of them are discussed elsewhere in this book. But a few points are worth making.

Most new products are produced and marketed either on a small scale initially, or, if on a large scale, then as an upgrading and extension of an established product in a developed market, as with distributed data-processing networks and now microcomputers. Such introductions are naturally made where a clientele of potential customers is already on the doorstep and where there is a keen awareness of the crowding both of competitive and complementary innovative activity forcing the pace.

There are two important qualifications to this general rule. The sale of new health care products made possible by advances in genetic engineering depends upon the incidence of the condition they are designed to treat. For example, the molecular cloning of a new vaccine for the treatment of parasitic diseases like malaria, hookworm or schistosomiasis would be almost wholly for use in poor countries. Development is dependent then on the second and third of the four factors listed above and not on the first. The second qualification is the rapid growth of joint ventures and licensing agreements whereby the technology is provided in the base of the innovating company and the selling tackled by companies with local marketing knowledge. This trend greatly widens the range of innovators and has recently extended Japanese leadership in applications of VLSI circuitry.[13] In the past, scale economies and the advantages of a large indigenous market led to an American preponderance. This has lately diminished and some of the fastest growing companies in the last ten years in data-processing have emerged in smaller countries,[14] apart from Japan.

Ronstadt and Kramer have recently pointed out that whereas in the late 1950s, 80 per cent of major innovations originated in the US, by 1965 this proportion had dropped to 55 per cent and is a good deal lower today. France has become a significant innovator in electric traction, nuclear power, aviation and the automobile industry; West Germany in chemicals, pharmaceuticals and heavy electrical machinery; Japan in opto-electronics, solid state devices, engineering and process metallurgy.[15] For all this, US-based multinational companies selling throughout the single global market, which—aside from products where language matters—is the relevant product market for the well-informed customers of high-technology products, retain a strong advantage. This advantage is enhanced by a growing interdependence between different scientific disciplines and different product, engineering and materials technologies in the development of families and systems of products. Telecommunications (drawing together aerospace, fibre optics, data-processing, hardware and software, modems and television) is the best-known current example. The symbiosis between different 'industries' in a single product market is beyond the reach of any but the largest individual economies. Convergent technology is increasingly likely in the future to destroy traditional industry classifications. Leaders in genetic engineering research products (such as enzymes), in large-scale fermentation, in immuno-diagnostics and in microprocessing are likely to be drawn together in the development, production and efficient delivery of health care products in the 1990s. And similar marriages of expertise will probably develop in education, energy and urban transport.

Successful innovating companies will increasingly be involved in collaborative activity with partners who may, for convenience, be located nearby. But, once such collaboration becomes an established pattern, many of the smaller contributors to the new technologies may be located in India, Brazil or Australia, for example. The absence of political risks, especially of restraints upon the movement of technology, management and capital, is likely to be the second most important determinant of which of the second-rank industrial countries will share the first fruits of innovative activity. The most important determinant will remain cultural and institutional: the will and ability of an economy to adapt quickly, transforming traditional skills, distributing the benefits of accelerated productivity so that the process is not resisted, and quickly taking to more advanced products as knowledgeable users at home and at work.

Notes and references

1. Schumpeter, J. (1947), *Capitalism, Socialism and Democracy*, London, Allen and Unwin, p. 84.
2. Schumpeter, J. (1934), *The Theory of Economic Development: an Inquiry into Profits, Capital, Credit, Interest and the Business Cycle*, Cambridge, Mass., Harvard University Press.
3. Schumpeter, op. cit., 1947.
4. Klein, L. and K. Ohkawa (eds) (1968) in *Economic Growth: the Japanese Experience Since the Meiji Era*, Proceedings of the Conference of the Japan Economic Research Center, Homewood, Illinois, observed that typically the growth of Japanese exports has been slower than the growth of her home market in those industries where her share of world exports has grown rapidly.
5. Gomulka, S. (1979), 'Increasing inefficiency versus low rate of technological change' in Beckerman, W. (ed.), *Slow Growth in Britain*, Oxford, Clarendon, p. 185 ff.
6. Marris, R. (1982), 'How much of the slow-down was catch-up?' in Matthews, R. (ed.), *Slower Growth in the Western World*, London, Heinemann, pp. 128–44.
7. Some of the work which developed this disequilibrium approach or applied it to the explanation of differences in industrial growth is: Schumpeter, J. (1947),*Capitalism,*

Socialism and Democracy, London, Allen and Unwin; Svennilson, I. (1954), *Growth and Stagnation in the European Economy*, Geneva, UNECEF; Lundberg, E. (1972), 'Productivity and structural change', *Economic Journal*, Supplement to 82: 465-85; Gomulka, S. (1979), 'Increasing inefficiency versus low rate of technological change' in Beckerman, W. (ed.), *Slow Growth in Britain*, Oxford, Clarendon; Stout, D. (1979), 'Capacity adjustment in a slowly-growing economy' in Beckerman (ed.), op. cit., pp. 103-17; Nabseth, L. and G. Ray (1974), *The Diffusion of New Industrial Processes*, Cambridge, Cambridge University Press; Blackaby, F. (ed.), *De-industrialisation*, London, Heinemann, pp. 56-77, 171-201, 263-86; Cornwall, J. (1977), *Modern Capitalism, Its Growth and Transformation*, London, Martin Robertson, esp. Chs 4, 5, 10 and 11.

8. See, for example, Carter, C. (ed.) (1981), *Industrial Policy and Innovation*, London, Heinemann, Chs 6, 7 and 8; OECD (1982), *Positive Adjustment Policies*, Paris; and Rothwell, R. and W. Zegveld (1981), *Industrial Innovation and Public Policy*, London, Frances Pinter, Chs 2, 5, 9 and 12.
9. See Mensch, G. (1979) on the depression accelerator effect in *Stalemate in Technology*, Cambridge, Cambridge University Press, and Stout, D. (1980), 'The impact of technology on economic growth in the 1980s', *Daedalus,* 109 (1): 159-67, esp. 160-2.
10. The problem is addressed by Nabseth in Nabseth, L. and G. Ray (eds) (1974), *The Diffusion of New Industrial Processes*, Cambridge, Cambridge University Press, Ch. 11. Many of the processes studied in the book seem to confuse symptoms and causes and times.
11. See Heertje, A. (1977), *Economics and Technical Change*, London, Weidenfeld and Nicolson, Ch. 10.
12. See Kennedy, C. (1964), 'Induced bias in innovation and the theory of distribution', *Economic Journal,* 74: 541-7.
13. Fujitsu is a prime example in joint ventures with SECOINSA (Spain) and Siemens (West Germany).
14. Norsk Data (Norway) and Nixdorf (Germany) are two of many examples.
15. Ronstadt, R. and R. Kramer (1982), 'Getting the most out of innovation', *Harvard Business Review,* 60 (2): 94-9.

12 Trade, technology transfer, and development
Meheroo Jussawalla

Introduction

Technology transfer has been a crucial issue of debate between the industrialised more developed countries (MDCs) and the industrialising less developed countries (LDCs). In the dialogue that has characterised international fora for the last two decades, technology and scientific innovation have been described as the common heritage of mankind to be equitably shared. This argument has reverberated through the deliberations of the New World Information/Communication Order (NWI/CO) as well as those of the New International Economic Order (NIEO). The technology gap and technological dependence have assumed political proportions that threaten to create distortions in the existing pattern of international trade. This problem is of equal significance to MDCs and to LDCs. As the industrial transformation of LDCs speeds forward, the technological comparative advantage currently enjoyed by the MDCs is eroded. Can new and innovative technology be continually renewed to prevent this erosion? On the other hand, LDCs fear the consequences of technology imports on their aspirations for self-reliant growth and sovereignty over resources. Specialised resources of skills and physical equipment are concentrated in the developed countries, which makes it cost-effective for LDCs to import their technology. While from the point of view of trade theory this is a rational approach, from the point of view of economic development, the perpetuation of the technology gap constitutes a new form of imperialism which the LDCs resent. Economic policies followed by them tend to distort market forces because of the use of shadow prices and opportunity costs of factor proportions. Market prices then do not reflect the actual scarcity of capital or the true opportunity cost of unskilled labour. Therefore, the constituents of any technology, indigenous or imported, may have a combination of several characteristics such as capital-intensity, skill-intensity, management-intensity, or maintenance-intensity.[1] This chapter looks at technology transfer in the context of international trade, the conflicts that emerge and the problems for developing countries which seem to defy solution. It deals with the costs, effects, and control of technology transfer as well as the opportunities for indigenous technology as an agent of development.

Technology transfer and trade theory

Technology transfer is not always channelled through trade flows because there are various economic influences that determine the content and trends of technology transfer and its application. Public capital in the form of economic and technical assistance has played a major role in transforming the social infrastructure of developing nations, because these countries have not been very successful in raising private portfolio capital in the markets of the rich countries. They have depended, to a considerable extent, on direct foreign investment for covering the gap in entrepreneurship, technology, marketing, and management. Such investment has also been viewed with suspicion as an exploitative mechanism by critics of international external aid programmes.

One of the major causes for the inability of LDCs to pay for their technology imports has been the emergence of protectionist measures in many industrialised countries. Primary products account for 34 per cent of LDCs' non-fuel exports and have been subject to wide price oscillations over the last ten years. In commodities other than primary products, the price gyrations have been as high as 15 per cent.[2] Simultaneously, the terms of trade for the developing countries have shown no appreciable gain. Despite the high value of the dollar in world exchange markets, the domestic economy of the US has faced unemployment, recession, and high interest rates. LDC exports have, in consequence, experienced a sharp deterioration in their terms of trade. All efforts at reducing protectionism in MDCs will fail so long as unemployment and recession continue to plague their domestic economies. The United States has drastically pruned its foreign aid allocations as well as its assistance to such institutions as the Export-Import Bank, the World Bank, The International Development Agency, and the International Monetary Fund. As enunciated at the Cancun meeting in October 1981, the United States' policy is to encourage Third World countries to borrow from the private sector. With this new direction in international aid, greater reliance for technology transfer is being placed on trade rather than aid.

However, trade-impeding policies continue to grow in the European Economic Community and in the US and Canada. There are a variety of restrictions against the traditional and diversified exports of LDCs. These measures, according to the *World Development Report*, have ranged from total prohibition of imports to tariffs, quotas, licensing of imports, voluntary export restraints, health and measurement standards, anti-dumping duties, countervailing levies, and orderly market-sharing arrangements. Such restrictions adversely influence the volume of exports of developing nations because they reduce the actual and potential comparative advantage which these nations enjoy and they hamper their efforts towards diversification of exports. In turn, these measures distort commercial and investment policies of the First World countries and bring about a less efficient global allocation of resources.

Traditional neoclassical economic theory stipulates the efficiency of resource allocation engendered by the spontaneous functioning of competitive markets, based on the Pareto-type assumption that initial distribution of assets and income is given and unchangeable. However, the literature of economics has developed a powerful case against markets as guarantors of efficiency even in the theory of pure competition.[3] This is especially true of markets in international trade, because they are creatures of political and social systems. Political decisions determine which markets can operate and which are to be repressed.[4]

Orthodox international trade theory affirms that economic relationships between countries depend upon the terms of trade and the patterns of exchange between them. The earliest versions of comparative advantage stipulated specialisation in exports for those commodities for which costs of production are relatively lower between trading partners. Heckscher and Ohlin improved upon this theory by postulating factor abundance and intensity as determinants of specialisation, so that free trade and specialisation will bring about factor-price equalisation. These theories assumed factor mobility across national borders, but trade restrictions widely prevalent today point to the breakdown of such trade theory.

Technical progress has an important bearing on the terms of trade because it influences the marginal productivity of both labour and capital. It can render the impact of new technology labour-saving or capital-saving or neutral when both factors have equal increases in marginal productivity. It is at this point of the analysis that technology transfer enters to alter the marginal productivity of labour and capital in developing countries, inducing exports of technology-intensive commodities from them. The probability of narrowing the income gap between the developed and developing countries will be greater if information-based technology transfer forms part of trade flows.[5]

Static assumptions of trade theory become untenable in a changing world of technical progress so that free trade does not maximise the welfare of less developed countries simply because it has no inherent tendency to equalise income and productivity levels.[6] The ratio of productivity to wages is higher in MDCs because information input and skill formation are higher. Financial capital tends to flow to those countries that have high productivity-to-wage ratios, thereby increasing the technology gap and the wage differentials. Consequently, deliberate policy formation becomes necessary to generate the type and quantum of technology transfer which will consciously build skill formation through aid, technical assistance, and preferential trade agreements.[7] Prebisch has challenged the assumption that technological changes induce higher productivity leading to reduced prices.[8] According to his analysis, prices fall only for the exports of periphery countries to the benefit of buyers at the centre. On the other hand, labour unions and monopolistic production systems prevent price reductions for exports from centre countries to benefit the buyers at the periphery. This is found to be true even for commodities produced within developing countries under the sphere of control of transnational corporations. If conventional wisdom is to be followed, then favourable terms for technology transfer should provide the best means for industrialisation of the Third World as well as an improved international division of labour that is better adjusted to factor endowments.[9]

However, in a real world situation, international trade theory and its impact on terms of trade fails to operate either on the basis of factor abundance or on the basis of availability, because of the sphere of control of transnational corporations (TNCs) that distorts the flow of benefits from technology transfer to the LDCs. Magdoff estimates that the internal capital accumulation of TNCs is generating output in host countries that exceeds the total volume of international trade.[10] This means that international trade is no longer the main vehicle of international exchange of goods and services. TNCs transmit centralised decisions to their subsidiaries, creating not only barriers to entry in the monopolised market system, but limiting technology diffusion by the use of exclusive networks. Every stage of production is covered by vertical integration.[11] In theory this pattern is treated as a new structure of international industrial organisation resulting in a new division of labour. Acquisition of bases in the developing world facilitates penetration of global markets by oligopolistic corporations, and Schiller claims that such penetration requires nothing less than total seizure of the host nation's information system and its ancillary parts.[12] Therefore, it is not technology alone but the organisation of the system of TNCs that imposes inequality.

International trade theory treats foreign investment as the export of surplus capital from industrialised countries seeking higher rates of return. This pattern is changing even for TNCs as high interest rates become more rigid in industrialised countries themselves. The dynamics of expansion are now tied not to high rates of return but to the organisational structure, so that opportunities for technology exports to LDCs are getting fewer. There is mutually penetrating investment among TNCs located in the First World. Access to transnational banking, to Eurodollars and Petrodollars, access to crucial data bases—all give TNCs a position of greater dominance in the flow of technology. Developing countries have to safeguard their balances of payments from adverse consequences of foreign capital repatriation, transfer of profits and dividends, and payments for imports of technology. Generally speaking, the hidden costs of technology transfer are high due to transfer pricing by TNCs which involves the over-invoicing of capital equipment imported by LDCs and under-invoicing of exports of the products of subsidiaries. Royalties and service fees add to the burden of imported technology on the balance of payments. Magee argues that TNCs deliberately use sophisticated technology in producing and transmitting information so that they can appropriate the returns on it.[13] This statement is supported by Schiller's analysis, in which he deals with the hegemony of TNCs in communication technology transfer *per se*.[14] He provides empirical evidence to show the control over global distribution of communications technology by American transnationals and the exorbitant costs of acquiring such technology. In other words, TNCs replace the market in trade theory and operate non-market systems of technology exports and planning. Therefore, in the real world, the theoretical conditions for Pareto optimality cannot be met. Flows of technology between subsidiaries of TNCs take place within closed rather than open markets.

Technology market imperfections are not only characterised by oligopoly but also by the information available to 'arm's-length' purchasers and the high costs of search. Bargaining for the purchase of technology by developing countries and the distribution of returns are important determinants of the process of technology transfer. The technology item to be purchased involves high costs of search. Quasi-rent becomes high because the marginal cost to the owner declines with further use of information about the technology in question.[15] Developing countries are learning to cope with these market imperfections by seeking reforms in the World Intellectual Property Organisation (WIPO) to enable competition in the technology market and to place obligations on those who own property rights for its use in developing countries. Klein argues that within the framework of a dynamic economic theory technological progress becomes more meaningful as it changes the impact of technology on trade patterns.[16] Randomness of communication is important to technological change and as trade patterns are altered by an open system of such change, policy for technology trade needs to be revised.

Another significant development in international trade is the concept of reciprocity being touted by the United States in an effort to reverse its current balance of trade deficit of $40 billion (of which $18 billion emerges from trade with Japan). In the name of free trade, it is a move to close US markets to those nations that impose restrictions on US exports.[17] The General Agreement on Tariffs and Trade (GATT), formed in 1947, provides a set of rules of international trade to its 86 member countries. But GATT has failed to keep pace with new problems of non-tariff barriers. The Tokyo round of negotiations ended in 1979 without any concrete policy for minimising non-tariff barriers. It does not cover international trade in services or investment in foreign countries and performance requirements now required by developing countries. The United States is attempting to close this gap in GATT regulations by introducing the principle of reciprocity and expanding it to cover services and investment. It is a particular form of protectionism and dangerous in its potential for trade war. In technology flows it is meaningless to expect reciprocity from developing countries who are confined by historical circumstances to be importers on a large scale. Whatever regulations they introduce to protect their foreign exchange balances are directed more towards the restrictive practices of TNCs and their appropriation of profits than at imposition of non-tariff barriers. If reciprocity is expected from LDCs it will amount to a further breakdown of the existing free trade system, even in its reduced form. GATT would also lose its influence over its members if the unconditional Most Favoured Nation principle collapses under bilateral agreements that reciprocity will involve.

The newest threat to technology transfer in international trade comes from what is termed 'countertrade' (a form of barter). These transactions, according to *Business Week*, constitute 25 to 30 per cent of total world trade as more and more governments attempt to narrow their balance of payment deficits.[18] For example, Brazil will permit car technology that is foreign-owned to import duty-free parts only if the car-makers export $21 billion worth of vehicles by 1989. Brazil has also required Canada's Spar Aerospace and Hughes Aircraft to arrange $130 million worth of Brazilian imports into Canada for a contract to build its space satellite. Such arrangements are a total breakdown of international trade theory as it has operated so far. It is tantamount to protectionism to support self-reliance. While the Soviet bloc has practised countertrade, it is now joined by OPEC and other First and Third World nations. It is an attempt to preserve employment within the bartering countries. If such agreements become mandatory and the free market closes down, technology exports from the MDCs may become fewer and adversely impact upon the development programmes of the developing economies. Can industrialised countries afford to export markets as substitutes for the export of credit? A solution to this crucial problem will determine the future of technology transfer within the framework of international countertrade. The international banking industry will receive a bonanza from such countertrade because it will need voluminous financing operations to export markets and it is likely that Third World countries will become more credit-worthy if the suppliers of technology contract to buy the output of the project. To put it simply, countertrade will generate long-term flows of hard currency earnings for the Third World provided First World corporations are willing to co-operate with countertrade. If they consider such an arrangement a threat to their profitability, which in effect it is, then countertrade may dry up the sources of technology transfer.

Robert Zimmerman argues that the entire contributing sectors in the balance of payments stand challenged by new informatics technologies that are replacing human workers throughout the world.[19] When computer programs regulate irrigation and robots work in motor-car factories, what chance do the growing populations of developing countries have of gaining employment through the transfer of technology under international trade arrangements? The intensive use of technology will increase productivity in LDCs but will simultaneously reduce their employment opportunities. If technology also ushers in new consumer and producer goods and services, it will create different kinds of employment opportunities requiring different types of skills and training for them. The advantage, perhaps a major one, of countertrade appears to be that foreign ownership of industry cannot be used exploitatively for the benefit of the owner who has not reinvested profits in the expansion of the local developing economies.

The question then arises: can trade in technology lead to development? In theory, free trade should reduce, if not eliminate, international differences in incomes paid to factors of production which in turn will lead to higher consumption and production in poorer countries. It will reduce the income gap as technology transfer further tends to equalise factor prices. These assumptions have had a pervasive influence on policy-makers who believe in trade expansion as a means of promoting development. But the continued differences in per capita incomes across nations has cast doubts about the relevance of trade theory, and, within its framework, the relevance of technology transfer for economic development. The differing ratios of productivity to wage rates have been an important factor in extending the technology gap, which only shows that if free trade fails to increase the capital endowment of developing countries, conscious decisions are needed to direct capital to LDCs in the form of technical flows and preferential trade agreements. That seems to be the major thrust of the new countertrade movement.

Does transfer of technology induce change?

It is widely anticipated that technology transfer will help developing societies to leapfrog from pre-industrial to post-industrial levels of transformation, but the empirical question remains whether technology transfer by itself can accelerate the pace of social change. To begin with, selective technology transfer has to be related to social justice because priorities have to be ascribed within limited domestic and foreign exchange resources.[20]

Transfer of technology induces social change in so far as the opportunity cost of the transfer is affordable for the recipient country. Developing countries are aware of the social system parameters within which technology is required to operate. Therefore, a development-oriented strategy of transfer has to rely on an optimal mix of technologies with varying capital intensities that are both capital-using and labour-using. For example, development planners are not overenthusiastic about total commitment to a 'small-is-beautiful' type village technology as being most suitable for reducing urban–rural differences. They would rather push for self-reliance in sophisticated science and technology and skill formation, and it was this view that was expressed by the LDCs at the UN Conference on Science and Technology held in Vienna in August 1979. Their demands shifted from an *en masse* transfer of know-how in the shortest possible time to building up their own expertise in research and development (R & D) and mobilising technology indigenously. Their programmes for Technical Cooperation Between Developing Countries (TCDC) and Economic Co-operation Between Developing Countries (ECDC) generated by UN organisations are aimed at collective self-reliance through exploring forward and backward linkages between oil-surplus countries and newly industrialising ones. In this way there will be a greater flow of south-south technology transfer and resulting social change.

In the fifties, an urban orientation to technology transfer led to the creation of enclave economies within developing countries so that the rural poor remained on the periphery, supplying their surplus output to the centre without any *quid pro quo*. Small urbanised pockets of prosperity had created a polarisation in decision-making, causing an equity crisis within LDCs. The urban elite found it in their interest to encourage multinational corporations to create islands of technology in backward societies and to exploit profits therefrom. Such a short-sighted policy in the fifties only perpetuated dependence on technology transferred from the First and Second Worlds. Despite the limitations of dependence, TNCs have continued for over two decades to enhance their operations in the developing world. They do this via intermediate economies, such as Hong Kong and Singapore, which feature common institutions with developed and developing countries.

As newer technologies replace human skills in many fields, the changes induced by their transfer will have far-reaching consequences on employment and its redistribution in developing countries. For example, the applications of informatics technologies are of the intensive type, which in the MDCs have led to office automation and the use of robotics. Given the present economic and demographic situation in LDCs, the use of informatics may generate productivity and economic growth, but will seriously damage employment levels. Historically, in industrialised countries, such displacement of labour in the long run has been evened out by increased productivity and the generation of new demand. Informatics has led to a tremendous growth in the services sector. But in a Third World without any cushion for unemployment compensation, the transfer of technology may be on the one hand cost-effective and highly productive, but on the other it may produce a more skewed distribution of income, at least in the short run. This is truer of societies where skill formation is already low and investment for retooling of skills is not readily available.

In two key industries, as Zimmerman reports, the change to high technology is already taking place in the Third World.[21] These are textile clothing manufacture and electronics components manufacture. The importation of new technologies in these industries is imposing a burden on foreign exchange resources. Having already attempted import substitution strategies that had to be abandoned, LDCs have erected protectionist barriers that impede the flow of new equipment. They find themselves trapped in a vicious circle of eroding productivity from labour-intensive processes and competitive outputs from international markets. In trying to preserve employment levels through labour-intensive low-cost technologies, developing countries are exposing their economies to market decisions by TNCs and continuing to fight a losing battle with poverty and underdevelopment. As Lester Brown affirms, the world is engaged in a historic transition towards the use of renewable energy sources, lower population growth, and more careful use of resources.[22] But employment, lifestyles, and industrial structures will be changing in a sustainable society, and are likely to alter the structure of the global economy. Third World countries will have to adapt to these changes to preserve their development prospects and survive in a highly competitive international environment. As the transition to such a sustainable society gathers momentum, new professions, such as agroforestry, will emerge along with new skills, such as skills to prevent soil erosion in agronomy. Informatics may provide some trade-offs to energy using transportation. Therefore, in the economic transformation that is anticipated, it is difficult to visualise an interdependent world in which the MDCs are technology-efficient and the LDCs are lagging behind. What makes the plight of the Third World worse is the fact that, despite technological abundance, the industrialised countries find their economies slowing down and their markets less affluent than they were in the past. Under these circumstances, paying for techology imports will become increasingly difficult for LDCs. In any analysis of the flow of technology across national boundaries, the factor endowments and differences arising therefrom are important determinants of transferring technologies from MDCs to LDCs, and their adaptation for location-specific development calls for further adjustments.

The most important problem that developing countries now face is the adjustment that is involved in getting the type of technology that will induce equity in income distribution. Is there a positive relationship between technology and equity? Ranis believes that there is, and developing countries in East Asia have been able to take advantage of technological progress with distribution equity.[23] It all depends on the role of technology transfer and the initial setting of the economy in which growth is planned and takes place. A flexible core technology, when selected by a developing country, gives room for ancillary technologies to be built around it and thereby ensures a wider distribution of disposable income. In this context, a flexible technology would not be in conflict with efficiency or factor endowments of the country. Technology choice, however, is limited by the fact that in the modern sector it originates in the developed countries. This is true even of agricultural technology, as witnessed by the Green Revolution with its varying degrees of success in different countries and even in different regions of the same country. However, the final choice and application of the technology is a mix of off-the-shelf transfer and its adaptation to domestic needs. For example, Taiwan and Korea made a more export-orientated technology choice compared with Mexico, Colombia and India which persevered with technology orientated towards import substitution. The LDCs which have in their process of development selected export-orientated technology have been able to combine the gains from growth with greater income equity. Taiwan was able to penetrate foreign markets by using technology that was sensitive to its factor endowments. Import substitution policies frequently sought changes in capital-intensive technology which distorted relative factor prices as well as relative commodity prices. Turnkey imports of technology follow such policies, resulting in greater inequity in domestic income distribution. Some economies have imported such technology into the public sector and sheltered that sector from competition. Rather than optimising returns on investment, the result has been satisficing economic behaviour by the larger production units.

Transfer of technology from the industrialised countries has enabled some developing countries to benefit from the advances without themselves incurring the costs of development. Transfer of high productivity techniques has, in some cases, led to concentration of wealth and technological dependence. Social welfare considerations have been overlooked and, according to a United Nations Industrial Development Organisation Report, it is the stratified system of international economic power relations that determines and reinforces technological dependence.[24]

During the 1970s new modes of international transmission of technology emerged due to many developments on the international economic scene. Some LDC governments took action ranging from outright expropriation of transnational assets to use of persuasion to influence their behaviour. The technologies and absorptive capacity of host countries has improved over the years. As a result, contractual and management arrangements have become more favourable to LDC development and self-reliance in technology products. Joint ventures are generally favoured by host countries, assuring that benefits will flow to the host country.[25] However, market imperfections and the internalisation of markets stand in the way of technology transfer to developing areas. Technology transfer needs back-up resources such as financial systems, organisation skills and managerial know-how. These are not readily available through the market. This makes it possible for technology to be supplied as a package and if the package is patented by the supplying firm, then monopoly rents and economies external to the technology are captured by the supplying firm. The experience of Third World countries receiving the technology has been mixed, depending on their ability to use and absorb the technology in the most efficient manner, which in turn depends on the availability of back-up resources. The extent to which technology can and does induce development in the LDCs will depend on the terms of the contract negotiated with the supplier. Dunning cites the example of computer components manufacture in which IBM has worked out an agreement with Argentina, Brazil, Chile, and Uruguay.[26] The components are manufactured in Brazil and assembled in Argentina, a Chilean firm is licensed to produce the punchcards, and the entire operation is controlled from Uruguay. The Ford Motor Company is planning to integrate all its plants in South Asia in a similar manner. Such modes of transmission do not have room for adaptation of the technology to in-country conditions. This process of adaptation depends on the country's educational system, the national ethos, and the types of government intervention. While LDCs enjoy 'latecomer' advantages, their ability to advance the frontiers of technology depends on government policy. Much of Japan's private sector has been able to acquire technological leadership chiefly because of the support it received from government.

Factor proportions and technology transfer

Technology transfer in developing countries has caused a problem of falling employment with high capital intensity. The processes and institutions creating new technology transfers often do not concern themselves with the high capital–labour ratios in manufacturing in LDCs. White projects the capital–labour ratios for new industries as $15,000 per worker and for petrochemicals as $200,000 per worker.[27] These high ratios pose serious economic and social problems. They increase the real wage difference between urban and rural workers, discourage the use of labour-intensive processes, and generate pressures on wages from government and labour unions. High capital–labour ratios have prevailed in LDCs on the grounds that they are efficient for increasing output and that other alternatives in the degree of capital intensity are inferior. Fixed proportions of factors have been considered superior to factor substitutability on the grounds that productivity is higher in the former and consequently there is greater efficiency. Galenson and Leibenstein argued that capital-intensive technology created surplus output which provided the scope for savings and ploughback.[28]

If a constant elasticity-of-substitution function is used to measure the percentage change in the capital–labour ratio in response to a change in the factor-price ratio, data on capital has first to be obtained. Empirical studies using econometric methods of regressing the logarithm of output per worker against the logarithm of wage, or regressing the output–capital ratio against the return to capital, all show that elasticities of substitution are mostly positive. In other words, efficient factor substitution is possible. Therefore, if wage rates are kept high and capital per unit of output is cheaper, then factor substitution will move towards greater capital intensity. However, economies of scale and optimal factor proportions provide alternatives to the choice of technology, allowing for a change from fixed factor proportions.

Newly industrialising countries, like Taiwan and Korea, have stretched existing capital by applying more labour-intensive techniques in textiles, electronics, and other industries. They have also applied labour-using techniques for such peripheral processes as materials handling and packaging. Kenya has used similar factor substitution in consumer goods manufacturing and packaging, but it is uncertain if the factor proportion flexibility yielded more efficient results in economic terms. If factor substitution faces internal and external competition for the output, then factor proportions have to become more and more efficient. This will depend on management efficiency and incentives for adaptation. Multinational corporations are generally reluctant to introduce factor-substitution in host countries, but there is varying evidence that MNCs operating in Mexico, Kenya and Korea have taken to factor substitution and adaptation of technology, depending on their management expertise.

Both Streeten and Stewart have found that wrong factor proportions are used in technology transfer because the wrong products are being marketed, such as cars, air conditioners, and refrigerators, which involve high capital-intensive technologies.[29] In turn, this is determined by the skewed income distribution in LDCs which enables the rich to control the production of consumer durables of the luxury type. The fundamental reason for the use of less appropriate technology is the scarcity of investment in R & D as an input into the process of development. The Schumpeterian argument was that the large absolute size of the firm and its market power are necessary to promote R & D because funds have to be generated for research. This hypothesis varies by industry; in electronics and the production of microprocessors, small size units may be responsible for R & D. However, the ready availability of cost-effective technology from MDCs and the scarcity of investment funds keep R & D in LDCs at very low levels. Apart from the benefits of 'latecomer' access to new technologies, developing countries do not incur considerable disinvestment in scrapping existing obsolete equipment. Many of them are able to start their process of modernisation with new imported technology. The only problem is that expertise in evaluating imported technology may be low, exposing LDCs to the risk of importing technologies with high rates of obsolescence.

However, LDCs do invest in national research institutes that undertake R & D and transmit results to domestic firms. Examples of such institutes are the Korean Institute of Science and Technology, the Indian Space Research Organisation, the Madras Leather Institute, and the Wheat Research Institute in Mexico. Such institutes are given investment funds more for applied research than for basic research since applied research is critical to the progress of industrial firms in developing countries.

Appropriate factor proportions are closely linked with the problem of indigenous technology development in LDCs. Stewart gives three reasons for promoting technological capacity in the Third World.[30] First, technological capacity is needed to adapt imported technology. Second, factor proportions in imported technology are generally not appropriate to LDC requirements, often leading to outputs of over-sophisticated high-income products. Third, continued reliance on imported technology results in a dependency relationship that may become permanent. Even with identical technologies there are marked differences in factor productivity in different countries. There may be differences in infrastructure, management expertise, and the skills of the workers. It is not always true that imported western technology leads to dependency and maldistribution of income. Japan's development in the twentieth century has been determined by the use of western technology, but Japan has spent one-third of its R & D investment on the adaptation of such technology, thereby generating higher productivity with minimal loss of employment. Essential to render imported technology efficient is the availability of adequate infrastructure: water, power, transport, finance, and management. When technology is imported with modern infrastructure, the productivity gains are high. Cost-efficiency also rises with competition in the market for technology because once a particular technology is developed, the marginal cost of its transfer falls. When exploring such markets, LDCs should make a thorough cost-benefit analysis before opting for technology transfer. Entrepreneurs make their own evaluation and banks analyse the economic viability of the transfer. UNIDO has provided guidelines for the evaluation of technology transfer agreements.

Costs of technology transfer

Generally speaking, developing economies are characterised by dualism, that is, a fast growing modern sector and a traditional sector that lags behind. The modernising sector faces a horizontal supply curve of labour. This means that with constant technology and capital costs, employment should grow in proportion to the rate of growth of output in the modern sector, which then expands without any changes in factor proportions, composition of output or technology. In theory, this is acceptable, but empirical evidence on poverty elimination and employment generation is not sufficiently strong in labour-abundant, less developed economies.

What is the role of technology in accounting for this failure? Prevailing factor prices attract producers to profit-maximising technology. Multinationals in quest of low-cost factors explore host country factor endowments. The LDCs, on their part, resort to prototype transfer in their quest for adaptation and indigenous technology development. Costs of transfer are reduced with improved infrastructure and stable economic policy in the recipient country. If a prototype is developed, the costs of the transfer would be further reduced if the operation of the new technology fits the needs of development. The reason for such cost reduction is that import-substituting technology used by LDCs has strong characteristics of system-maintenance grounded in market dynamics. For instance, costs of labour-absorbing technology may be higher depending on changes in goal emphases, foreign exchange availability, and task environment.

Costs of technology transfer cover costs of transmission and absorption. Both are high when the technology transferred is complex. This is true because the costs of resources utilised to effect the transfer plus costs of know-how, royalty, or quasi-rent for securing access to technology have all to be accounted for. Peripheral support costs are necessary for generating information flows about the technology choice. While computing the economic costs of technology transfer between nations, an important variable is the type of technology imported. If leading-edge technology is transferred, the costs will be higher because its success or failure has not been adequately tested in the short time span in which it has operated. On the other hand, state-of-the-art technology will be less expensive since it has been understood and used over a long time span. Competition is also greater for the supply of the latter type, resulting in cost-cutting. In addition to import costs, the costs of planning, establishing and operating the transferred technology become relevant.

Financing the transfer is also a variable in costs. Is the transfer effected through official government agencies or by direct foreign investment? There are also hidden costs and intangible benefits that are not easily measurable. To the recipient country, the costs become worthwhile if the technology transfer serves to cover the gap between net domestic savings and projected investment. There is a benefit if the technology transfer supplements local capital which is used for generating a higher GNP. The costs will be related to the flow of profits to the foreign investor and royalty fees, as well as the costs of diverting resources from domestic firms engaged in similar production of commodities and services, but not using the same high degree of production technology. The social cost is more serious in so far as it dampens the initiative for local R & D which, in turn, prolongs dependency on external sources. In such cost evaluation, we find that the economic feasibility of the transfer is more important to the developing country than the technical feasibility.

The concept of a production function, as it obtains in neoclassical economic theory, is not pertinent to the dynamics of technological change in developing countries, because intermediate technology does not create the same substitution effect of using more labour or more capital as it does in advanced economies. The possibilities of factor substitution emerge from technological innovations accomplished in the past. This is not available to many developing countries. Often technology is transferred through different channels and its spin-off effects are distinct from factor substitution. Consequently, focus has to be shifted from quantifiable costs to socioeconomic costs which are borne by the recipient country. This is even more applicable because the choice of a specific technology for development planning is influenced by the socio-economic environment into which it is transferred, and cannot be written off as disembodied knowledge. The real welfare effects will depend on the conditions in which the transfer is accomplished. Economic costs will vary in accordance with the monopoly power of the supplier and the absorptive capacity of the buyer.

Developing countries are willing to bear the high opportunity costs of technology transfer because they want to accelerate their entry into the high-technology high-growth era. The choice of technology is crucial, not only because of its costs and effects on the production system, but because of the value content. Dramatic transitions in societies can take place through technologies like those of satellites, computers, and instructional television. These may give rise to a conflict in value systems which may need the formulation of new societal goals.

If investment in technology is geared to the satisfaction of the basic needs of rural people in developing areas, then the selection of technology has to be problem-specific and location-specific. There can be no particular technological fix or a pure technology solution to the problems of poverty and underdevelopment. Technological change is tied to preferred outcomes which form part of the objectives of planning development. Therefore the costs have to be measured in the context of the social order. For example, the People's Republic of China has used technological autonomy in a closed self-contained approach to development since 1960, whereas Taiwan and South Korea have been predominantly open, using technology for exports as the leading sector for development. On the other hand, Brazil, Mexico and India have imported technology for industrialisation but have failed to meet the problems of income distribution and poverty of the peripheral people.

The process of technology transfer involves more than the creation of research institutes and national technology centres, improved access to foreign patents, and the availability of finance to exploit them. The appropriation of knowledge is the mainspring of innovation.[31] This requires scarce resources of skill that have a high opportunity cost. It requires the capacity for science and technology to be linked with the production and education systems in order to engender innovation. Not only are there financial and skill restraints in LDCs for technological innovation, but the commercialisation of research findings is also comparatively low in these countries. In other words, competition from domestic markets for technological innovation being limited, imported technology can be priced high, depending on import elasticity of demand.

Costs of technology transfer are closely linked to external deficits and international debt. It is estimated that 40 per cent of total exports of capital goods and equipment from OECD countries go to the developing countries.[32] However, balance of payment problems have aggravated the restraints on such exports, even for some OPEC countries which are reducing their modernisation schemes to stem the adverse impact on international payments. As deficits mount, pressures increase on LDCs to export manufactured goods and construction and other services to cover deficits. Devaluation as a tool for increasing exports is rendered less worthwhile if the external deficit is large and persistent. This is because a developing country cannot afford to spend too much time in the pit of the J-curve.

Among developing countries there are wide differences for financing technology transfer. For example, Mexico and Argentina (before the Falklands War) have no real constraints from trade balances. Brazil has large commodity exports, but still has constraints arising from external deficits. Taiwan is among the Newly Industrialising Countries (NICs) that have continued surpluses on visible trade account and can finance technology imports. Korea faces problems with current account deficits and a declining growth rate. Hong Kong and Singapore have their exports orientated towards labour-intensive production, though their role as banking centres eases the constraints imposed by external debt on technology transfer. It is for the poorest developing countries that oil imports and external debts are crucial constraints on investment in technology transfer. They are unable to tap commercial international markets as the NICs have done. They have to rely on foreign aid to meet the costs of technology transfer.

Effects of technology transfer

It is not possible to claim that transfer of technology will by itself generate economic development. Diffusion of technological change can be gradual or sudden.[33] It is possible for a new technology to remain unused until factors conducive to its adoption and diffusion force the volume of demand to expand. *The Economist* points out that, for the first time in a hundred years, America's lead in technology is being challenged by Japan.[34] In electronics Japan is already drawing level with the United States. What is impressive is how Japan started with the process of technology transfer and imitative products: today industrialised countries clamour for restraints on Japanese exports of cars, television sets, and microprocessors. Japan's next export industry, according to *The Economist*, is technology.

The total effects of technology transfer are difficult to isolate and measure. At the microeconomic level of the firm there is clearly an advantage, otherwise the transfer would not take place. In order to maximise these benefits from transfers at the macro-level, a great deal of restructuring of world industry will become necessary; perhaps a new international division of labour will be needed. There is conflicting evidence about the positive effects of technology transfer at the macro-level. Growth stemming from technology transfer has been most evident in Latin America and South-East and East Asia. These industrial growth centres have emerged to change the equilibrium of existing industrial flows across countries. These changes are evident in the trade flows between industrialised and developing countries. The gross or direct effects are those that result in the establishment of new competitive industrial capacity. The net effects are those that cover new trade patterns, prices, and income effects. Some sectors will be stimulated by technology transfers, while others will lag behind. The net effects will also vary between countries, depending on such factors as a country's cyclical, structural, and social conditions. For example, for the transferor the positive effects may be sales of technology and equipment, and low-risk investment in the short term. The long-term positive effects may be the opening of new effective demand, production of sophisticated techniques and products, lower prices for them, and job creation. But the negative effects in the developing countries may be resource exploitation, deficits in payment balances, dependency for R & D, high costs of adaptation, and high opportunity costs.

OECD imports from LDCs have increased at an annual rate of 28.9 per cent from 1970 to 1975, but dropped to 26 per cent in 1979. Similarly, OECD exports to LDCs rose by 23.8 per cent between 1970 and 1975, and dropped to 12 per cent in 1979. After 1973 most OECD exports flowed to oil-producing countries, but the slowdown in economic activity in the OECD countries made developing countries buy more capital equipment. In 1980 about 40 per cent of capital goods exports from OECD countries were purchased by developing countries, placing their trade balances in deficit. Globally, the picture that emerges is one of trade being stimulated by technology transfer. Figure 12.1 shows the growth of OECD imports from LDCs between 1970 and 1979. The United States and Japan have been the biggest buyers of developing country exports of manufactured goods. There has been an industrial restructuring process for which technology transfer has been both a cause and an effect. Export processing zones and south-south trade have been some of the spin-offs of technology transfer.

A global system of technical and economic co-operation among developing countries is emerging as one of the major impacts of technology transfer. LDCs realise that they need to strengthen their technology capacities by becoming active agents of their own transformation. Self-reliance as a goal was enunciated by them at the meeting of nonaligned countries at Cocoyo (Mexico) in 1974. The pursuit of increased self-reliance did not exclude technical co-operation with developing countries in order to strengthen their collective technological capabilities. Consequently, a great deal of attention is focused on ECDC and TCDC, which constitute an important framework for regional and interregional co-operation for trade and technology. Under these programmes of technical and economic co-operation, the adverse effects of technology imports via TNCs are considerably reduced because existing technologies within LDCs are strengthened and communications among LDCs are improved. Indirectly such programmes expand international co-operation. The sectors in which such south-south co-operation becomes feasible include a wide range of consumer durables, intermediate products, light and medium engineering goods, and machinery—products with which many NICs are competing in international markets. Maintenance, engineering, and consultancy services, and joint ventures are also acquired. However, the oil-exporting countries still prefer to import the most modern and sophisticated technology from OECD countries and refrain from entering into agreements with LDCs for such imports. UNIDO proposes to function as a channel for the joint acquisition of technology by developing countries. This could be achieved through a process of collective bargaining since development plans in LDCs are more or less similar, concentrating on steel, petroleum, fertilisers, textiles, sugar, cement and agro-industries.

The Brandt Commission has emphasised the purpose of economic co-operation

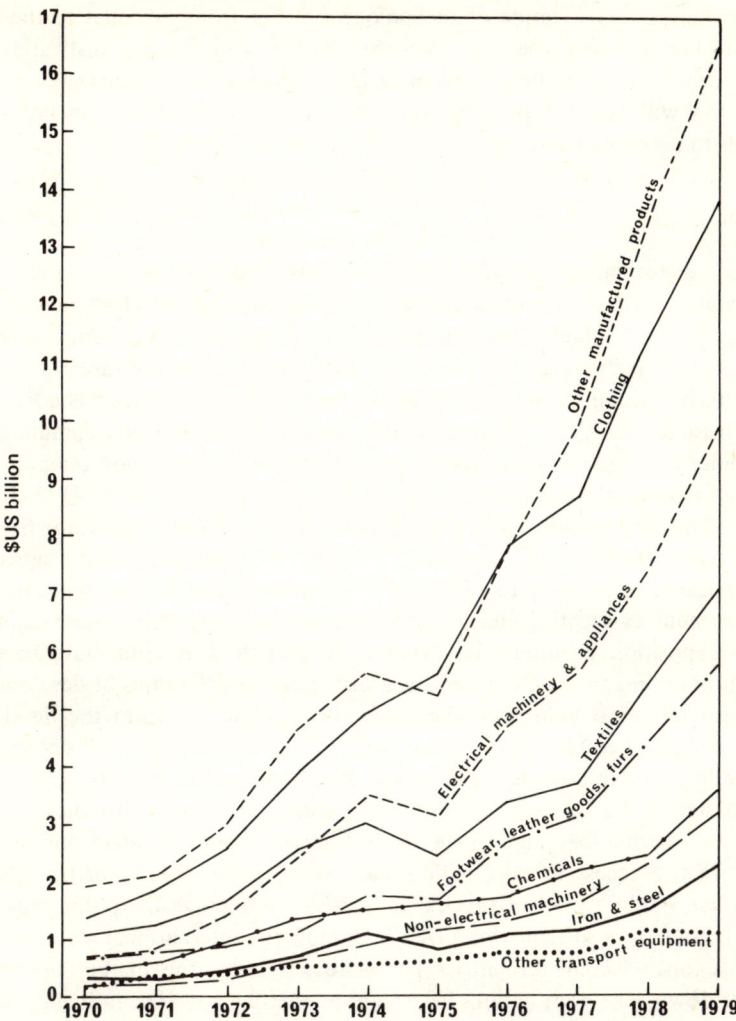

Source: OECD (1981), *North-South Technology Transfer: The Adjustments Ahead*, Paris.

Figure 12.1. OECD imports of manufactured products from developing countries by broad product groups, 1970-79

among developing countries and the need to exploit their potential for economic and social development.[35] The Caracas Conference in 1981, which endorsed this view, outlined finance, transfer of technology, and trade as important sectors for co-operation. The Caracas programme specified a solution to the problem of upgrading the technical capabilities of LDCs by exchange of experience as well as seeking improved terms of technology transfer from the First to the Third World. For this purpose multinational research and training institutions are proposed to speed up the process of TCDC. Trade preferences among developing countries will be an important effect if technology transfer at an intra-south level is to become operative.

Conclusions

Buyers of technology in LDCs are faced with imperfect markets and limited information. They are not conversant with all the options open to them and alternative technologies relevant to their specific needs. Consumer power is restricted in technology markets and LDCs have to invest in ascertaining the legal, technical, and financial information pertaining to contracts. The MacBride Commission has suggested impact statements for the transfer of communications technology in order to control the adverse effects on the socioeconomic structures of developing countries.[36]

Whether technology transfer is a parameter or a problem generator has been discussed. Industrialised countries question the suitability of transferring sophisticated technology to LDCs on the grounds that it works against national development by creating and strengthening elite enclaves within those countries. Such a position is untenable. Technology growth does contribute to social transformation and promote improved distribution of the gains of development. It seems unfair to hold back the benefits of technology from the developing world simply because elite groups are, at present, powerful. If demand for better living conditions is an indicator, the peripheral people are conscious of the existence of a 'glass curtain economy' and want the good life they see lying behind it. Technology may prove the leveller which reduces elitism and provides cost-effective access to information and science systems. It is for the TNCs in the First World to make specific affordable technological options open for LDC buyers. It is difficult for LDCs to exploit the international market system because once income disparities penetrate a system, the market mechanism breaks down. The poor nations have scant purchasing power to influence market decisions, but what are the ways of speeding up technology transfer on a scale hitherto unknown? Much will depend on the commitment and depth of interest of decision-makers from both the industrialised and the developing nations. The challenge to international policy lies in regulating the flow of technology without impeding its flexibility. Eric Eckholm puts the matter succinctly when he writes, 'the absolute poor have no time to worry about global trends'.[37] The key idea seems to be sustainability of the world's poor majority. What we need is not only development, but the right kind of development with a just social order. Development that is not sustainable will destroy the very resource base on which it ultimately depends.

In a foreword to the book *Down to Earth*, Barbara Ward wrote with a strain of pessimism: 'The last few years have seen a rising resentment of the notion that we are all dependent on other people and other nations . . . Big powers never want to be dictated to by little ones'.[38] Perhaps attitudes will change as global interdependence becomes more real. A vision of joint survival will lead to a better sharing of science and technology and to fuller co-operation in exploiting the planetary resources for the benefit of the poorer nations. It may become possible in the near future to transfer technology without adverse effects. Lamberton recommends 'technology without machines',[39] or organisation as a method for transcending limitations imposed by complexity. New organisational forms may make the future possible for technology transfer to generate development of periphery countries, and periphery people within them.

Notes and references

1. Robinson, A. (1979), 'The availability of appropriate technologies' in Robinson, A. (ed.), *Appropriate Technologies for Third World Development*, New York, St Martin's Press, pp. 26-31.
2. World Bank (1981), *World Development Report 1981*, Washington D.C.
3. Helleiner, G. (1978), *World Market Imperfections and the Developing Countries*, Occasional Paper No. 2, Overseas Development Council, Washington D.C., US Government Printing Office.
4. Diaz-Alejandro, C. (1976), 'North–South relations: the economic component', *International Organisation*, **29** (1): 213-41.
5. Jussawalla, M. (1982), 'International trade theory and communications' in Jussawalla, M. and D. Lamberton (eds), *Communication Economics and Development*, New York, Pergamon.
6. Kitamura, H. (1968), 'Capital accumulation and theory of international trade' in Livingstone, I. (ed.), *Economic Policy and Development*, Harmondsworth, Penguin.
7. Colman, D. and F. Nixson (1978), *Economics of Change in Less Developed Countries*, New York, John Wiley and Sons.
8. Prebisch, R. (1959), 'Commercial policy in underdeveloped countries', *American Economic Review*, **49** (2): 251-73.
9. Amin, S. (1979), 'NIEO: How to put Third World surpluses to effective use', *Third World Quarterly*, **1** (1): 65-72.
10. Magdoff, M. (1976), 'Multinational corporations and social development' in Apter, D. and L. Goodman (eds), *Multinational Corporations and Social Change*, New York, Praeger, Ch. 8.
11. Helleiner, op. cit.
12. Schiller, H. (1979), *Communication Accompanies Capital Flows*, Document No. 47, prepared for the International Commission for Communications, Paris, UNESCO.
13. Magee, S. (1978), 'Information and multinational corporations' in Bhagwati, J. (ed.), *New International Economic Order*, Cambridge, Mass., M.I.T. Press, pp. 318-37.
14. Schiller, H. (1981), *Who Knows? Information in the Age of Fortune 500*, Norwood, New Jersey, Ablex.
15. See *Bulletin of the Institute of Development Studies*, University of Sussex, **3** (1): 16-23.
16. Klein, B. (1978), *Dynamic Economics*, Cambridge, Mass., Harvard University Press, pp. 164-6.
17. Kirkland, R. (1982), 'Washington's trade: war of words', *Fortune*, **105** (7): 35-9.
18. 'New restrictions on world trade', *Business Week*, 19 July 1982: 128-32.
19. Zimmerman, R. (1982), 'Technological innovation and economic development', *Agora*, January–March: 28-31.

20. Jussawalla, M. (1980), 'The impact of communication technology transfer on economic development', *Telecommunications Policy*, 4 (4): 249-62.
21. Zimmerman, op. cit.
22. Brown, L. (1982), 'Living and working in a sustainable society', *Futurist*, 16 (2): 67-70.
23. Ranis, G. (1981), 'Technology choice and the distribution of income', *Annals of the American Academy of Political and Social Sciences*, 458: 41-53.
24. United Nations Industrial Development Organisation (1979), *Technological Self-Reliance in the Developing Countries*, Series No. 15, Vienna, UNIDO-ID/262, pp. 9-10.
25. Dunning, J. (1981), 'Alternative channels and modes of international resource transmission' in Sagafi-Nejad, T. *et al.* (eds), *Controlling International Technology Transfer*, New York, Pergamon.
26. Dunning, J. (1977), 'Trade, location of economic activity and the MNE: a search for an eclectic approach' in Ohlin, B., P. Hesselborn and P. Wijkmon, *The International Allocation of Economic Activity*, London, Macmillan.
27. White, L. (1979), 'Appropriate factor proportions for manufacturing in less developed countries' in Robinson, A. (ed.), *Appropriate Technologies for Third World Development*, New York, St Martin's Press, pp. 301-29.
28. Galenson, W. and H. Leibenstein (1955), 'Industrial criteria, productivity and economic development', *Quarterly Journal of Economics*, 69 (3): 343.
29. Streeten, P. (1972), *The Frontiers of Development Studies*, New York, John Wiley and Sons, pp. 223-38; Stewart, F. (1973), 'Economic development and labour use', *World Development*, 1 (12): 25-8.
30. Stewart, F. (1981), 'Arguments for the generation of technology by less developed countries', *Annals of the American Academy of Political and Social Sciences*, 458: 97-109.
31. See United Nations Industrial Development Organisation, op. cit.
32. OECD (1981), *North-South Technology Transfer: The Adjustments Ahead*, Paris.
33. Mandeville, T., S. Macdonald and D. Lamberton (1980), 'The fortune teller's new clothes', *Search*, 11 (1/2): 14-17.
34. 'Japanese technology survey', *Economist*, 19 June 1982: 64ff.
35. Brandt, W. (1980), *North-South: A Program for Survival; Report for the Independent Commission on International Development Issues*, Cambridge, Mass., M.I.T. Press.
36. MacBride, S. (1980), *Many Voices, One World*, London, Kogan Page/UNESCO, Part V.
37. Eckholm, E. (1982), *Down to Earth*, cited by J. Wiley in 'Phenomena, comment and notes', *Smithsonian*, June.
38. Ward, B. (1982), in Eckholm, E., op. cit.
39. Lamberton, D. (1981), 'Technology without machines', paper presented to the R. S. Gynther Conference on the Impact of Technological Change on Organisation, University of Queensland, Australia, October.

13 The technology transfer process in foreign licensing arrangements

Lawrence S. Welch

Introduction

Technology transfer between nations takes place in a variety of forms and through a variety of business arrangements. Very often the technology is transferred in the 'embodied' form of physical goods, or as part of a significantly wider arrangement, for example, as an element of the foreign investment package or a systems sale. While much of the debate surrounding technology transfer, particularly from the advanced to the developing countries, has focused on the appropriateness of the technology for the recipient nation, and the price and conditions of transfer, less attention has been given to the efficiency and effectiveness of the technology transfer process itself.[1] Although it has sometimes been analytically convenient to depict technology as a good, this has led to an oversimplification of the demands of the transfer process.[2] Technology is not simply purchased and transferred in an 'off-the-shelf' manner; frequently a complex process of definition, marketing, negotiation and implantation is involved over an extended period. 'It is much more accurate to view technology transfer as a relationship rather than as an act.'[3] The transfer may be part of a multi-faceted interaction between two organisations over time.

In this chapter the international technology transfer process is examined from the perspective of licensing arrangements. From an aggregate perspective, foreign licensing occurs principally between related organisations. For example, in 1976 the proportion of total US receipts from royalties and management fees which were intra-firm had reached 82 per cent.[4] Other countries appear to be following the US pattern although the intra-company proportion tends to be lower because outward foreign investment activities have not yet developed as strongly. The intra-company proportion of total Australian receipts from the provision of technical know-how to foreign organisations was 45 per cent in 1976-77.[5] Within the foreign investment framework, though, licensing appears as only one element of the overall strategy, and its purpose is often connected with considerations such as control, the transfer of funds, and taxation reduction.[6] Thus, licensing, as a medium of technology transfer, will be examined outside the foreign investment context in this article—as a relationship between independent organisations.

While it has been argued that the marginal cost of transferring new technological information is low once the development costs have been incurred, empirical evidence indicates that transfer costs can be considerable.[7] They may even be such as to act as a real impediment to the effectiveness of the technology transfer process.[8] In fact, transfer costs and a variety of constraints emerge at different stages of the transfer process in licensing, and it is an important objective of this chapter to assess their nature and significance in determining the effectiveness of the whole process. Technology transfer is taken to refer to the whole range of activities by which licensor and licensee come into contact, negotiate an arrangement and carry through the transfer demands of the arrangement. The object of the exchange will, however, frequently encompass more than just technology.

The licensing package

One of the reasons why licensing is often favoured as a means of technology transfer is because it allows a company to purchase the required technology without the implications of foreign ownership. Japan is frequently cited as an example of a nation that has been able to restrict foreign investment but still obtain the technology required for industrialisation by a deliberate policy of licensing. By so doing, Japan has been able to 'unbundle' the foreign investment package and extract the parts which were most appropriate to its own situation. This policy was clearly assisted by Japan's heavy emphasis on technological absorption in its research and development as a means of ensuring the utilisation of the technology transferred.[9] Thus, licensing appears as an attractive, low-cost means of isolating and purchasing the specific technological component required.

Nevertheless, licensing normally implies more than a transfer of 'pure' technological information. While the component parts of the technology transfer in the licensing arrangement are more clearly defined, it is nevertheless true that, like foreign investment, there has been an inevitable building up of a 'package' around the technological core. One important element of this broader package is what Stewart has defined as 'marketing rights':

> Technology was defined as knowledge of how to do and make useful things. But in practice examination of the market for technology suggests that in the process of commercialization of this market, the content of technology transfer has become more complex than this. A major element in technology transfer is the acquisition of the right to use certain trademarks and/or access to certain markets and inputs. For shorthand we may describe the acquisition of trademarks, and privileged access gained to markets and/or to inputs as *marketing rights*. These may be highly valuable to individual firms in helping gain markets or inputs . . . the acquisition of marketing rights forms an

important element of *costs* and is also a significant aspect of *motivation*. Discussion of the international transfer of technology thus covers both the communication (or sale) of knowledge and the sale of marketing rights.[10]

The gradual evolution of the broadening licensing package represents a response to pressures on both sides of the licensing arrangement. For the user there is a concern to ensure that the technology is translated as rapidly as possible into an efficiently working, and marketable, form. To achieve this, relevant working knowledge, and any associated rights, which can help to assure the outcome of the transfer process, are clearly desired. The line between technological and other forms of knowledge in the transfer process is blurred from the user's perspective: he is seeking all of those interrelated elements which add up to a perceived likelihood of commercial viability.

From the licensor's perspective there is an inevitable concern to wrap the technological information, which may be the basis of a transfer, into a more secure package. By broadening the elements in the package into such areas as marketing and managerial know-how, and marketing rights, the whole is made much stronger from the point of view of protecting the firm's proprietary rights over its industrial property. Not only that, but the marketability of the package is improved and the basis of income generation is extended. The broadened package provides a better basis for assuring returns in the short run and market development in the long run.

The pressures from both ends, therefore, have strengthened the trend towards a broadening of the licensing package, with growing emphasis on commercial know-how and a more complex intertwining generally of the technological and commercial elements. Some indication of the growth in importance of 'commercial technology' and associated rights is given in a recent study of Finnish industrial companies licensing to independent foreign licensees. The proportionate inclusion of the different objects of licensing in licensing arrangements was was follows:[11]

technical know-how	96.1%	marketing know-how	24.7%
patents	48.0%	management know-how	11.7%
trademarks	36.4%	designs	5.2%

As well as a broadening in the objects of licensing contained in the package, licensing arrangements may also be extended in a number of other ways. Licensing is frequently associated with a variety of exports to the licensee, such as plant and equipment, component parts or raw materials. The aggregate value of exports of associated plant and inputs by Finnish companies to independent licensees was more than double the value of licensing income.[12] In a study of Australian companies licensing abroad, just over one-third of respondent firms reported that the arrangement had resulted in sales of associated products.[13] The licensing relationship may also be extended into other areas of co-operation, for example, cross-licensing or joint activity in third markets. Such cases will often emerge as the relationship evolves over time.

Thus, the licensing package, and licensing relationship generally, which has been evolving as a medium of technology transfer, has tended to become broader and more complex around the technological core. This has had the effect of making licensing a more complex exercise, and of increasing the demands on both licensor and licensee in the whole transfer exercise—to identify and assess relevant company knowledge, to undertake the necessary registration of industrial property, to develop the various elements into a cohesive whole, to undertake negotiations across a wider spectrum, to effect the range of transfers required and to develop a more comprehensive interaction between the two organisations as a consequence. In general, a broader 'systems' approach to licensing appears to be demanded.

Patents and know-how

As the above Finnish evidence indicates, patents continue to occupy an important position within the licensing package, although apparently less important than unpatented know-how. Helleiner has argued that: 'the consensus emerging seems to be that unpatentable knowledge with respect to the process of production is of greater significance than patented know-how . . . Knowledge embodied in the patent is, in any case, normally insufficient, by itself, to permit its efficient working.'[14] This pattern was confirmed in an Australian study which indicated that, of a sample of mainly smaller companies licensing overseas, 32.6 per cent had not attempted to file for overseas patents. The overall breakdown showed that patents were responsible for only 20.1 per cent of licensing receipts, compared with a figure of 69.8 per cent for the know-how component.[15] While there was considerable variation on a company-to-company basis, the general impression gained was that unpatented know-how was recognised to be the crucial element in the technology transfer process—by both licensors and licensees. The pre-eminence of unpatented know-how demonstrates that the clearly specified technical information for public registration does not fulfil the demand of effective technology transfer in most situations. The technological know-how which the companies consider of greater importance is of a more intangible, company-specific nature, and requires person-to-person interaction for the transfer to be realised.

Ability of the technology recipient

Ability to use bare, documented technological information is, of course, partly dependent on the technological absorption capacity of the recipient. In general it can be argued that the wider the gap between the technical skills of the transferor and the recipient, and the more complex the technology in question, the greater the demand on the companies to bridge the gap in the transfer situation. At the extreme, the less developed countries, with limited absorption ability, are likely to place even greater reliance on non-patented knowledge to assure effective transfer. Studies by Contractor indicate that less developed countries place greater emphasis on organisational and production management assistance in licensing arrangements than do advanced countries.[16] He notes that, 'technologically advanced firms will frequently obtain bare patent rights and go on to produce on their own, because they have already "internalized" knowledge from past experience with similar products'—although this did not happen in the majority of cases.[17] Absorption ability tends to be related to specific technical skills within the company, though, and the further it has to move away from the skills foundation it possesses, the greater the need for extended transfers of technological information and assistance in the licensing arrangement—even in technically advanced companies.

Importance of patents

In some cases, companies are unprepared to take out patents because they have a poor view of their value. Patents may be considered to provide insufficient protection against direct infringement, or regarded as being too easy to invent around. Nevertheless, they remain the principal public means of establishing proprietary rights to the firm's technology, and, in fact, appear to play a far broader role in the technology transfer process than simply industrial property protection and definition of the firm's core technology.

Patents are an important consideration for the licensee—in providing a degree of protection in the licensee's market as well as perhaps forestalling potential competition. They also provide some measure of the licensor's technological credibility to an uncertain licensee. When the knowledge gap between potential licensor and licensee is large, patents represent one important means of helping to bridge it, and of reducing uncertainty. In an Australian study, it was reported that patents were particularly necessary as a prelude to licensing into the US market: 'many US companies regarded patents as a guarantee or safeguard of the technical respectability of the licensor and would not enter into negotiations without them'.[18] In addition to their individual role, however, patents make a vital contribution to supporting the total licensing package, and making it more marketable. This strengthens the hand of the licensor in negotiations. The same Australian study found that larger companies, and those with greater international licensing experience, tended to have a more positive view of the value of patents.[19] This appeared to be because of a clearer recognition of the marketing value of patents. In one company it was interesting to note that international patenting and licensing activity had been initiated by the marketing manager, despite the scepticism of technically orientated executives, who considered patents to be of little value.

Inventors without a manufacturing base, and without the assurance of a broader licensing package, clearly tend to view patents differently from manufacturers. The patent may be the only source of income-generation. They, therefore, tend to rely heavily on the abilities of the licensee to recognise the value of the technological information and to apply it within a production and marketing context. The exposed position of the inventor weakens his ability to carry through the marketing and negotiation exercise. The problems experienced by inventors in selling and transmitting their unique knowledge give an indication of the limitations of licensing only the bare technological information if the technology transfer process is to operate effectively.

Other package components

The unpatented know-how component of the licensing package is normally the crucial element of effective technology transfer because it is the information which enables the technology to be made to work in practice. For the licensor, possession of such know-how strengthens his bargaining position and makes him less susceptible to patent infringement, because the patent of itself is usually insufficient to demonstrate how the technology operates in practice. Considerable costs have normally been incurred—in the areas of development, manufacturing start-up and commercialisation—in the process of generating the firm's unique know-how. These represent a significant barrier to those companies which merely have patent details as a starting point. From the licensee's standpoint, of course, such costs are one of the reasons why the know-how factor is valued so highly—it allows production to occur without repeating many of the learning costs associated with development. In addition, it allows the licensee to proceed more rapidly to the marketplace. In this latter respect, managerial and marketing know-how may be just as important as technical know-how. Because much of the know-how is intangible in character, and firm-specific, the demands on interaction between the two parties in order to effect the transfer are increased, while the preceding negotiation process becomes more complex and difficult. The value and effectiveness of the know-how are difficult for the licensee to assess in the negotiation situation.

Trademarks, designs and other marketing rights, such as access to certain key inputs, are also valuable related elements of the licensing package. Trademarks particularly have tended to be stressed because of their contribution to marketing penetration. While not in themselves communicating useful knowledge, they may be important in the marketing build-up, and in providing a further element of protection for the licensing package.

Each element of the licensing package, therefore, contributes to the range of knowledge and rights which can potentially assist the licensee in establishing and commercialising the technology in question. The packaging of technology and associated rights under licensing is becoming more complex and the transfers more demanding, but in part this is a reflection of the complexity of the technology transfer process itself.

Exchange demands

The effective transfer of the various elements of the licensing package is a demanding process, involving a variety of transfer mechanisms, flows and types of interaction between organisations and personnel. These activities are reflected in the range of transfer costs. Contrary to the notion that the marginal cost of transferring technology, once developed, will be low, the evidence indicates that transfer costs can be very substantial. For example, in a study of the cost of technology transfer by US multinationals, including both transmission and absorption costs, Teece found that transfer costs were on average 19 per cent of total project costs, ranging from 2 per cent to 59 per cent.[20] Other studies, specifically of licensing, confirm the importance of transfer costs.[21] Oravainen's study included an assessment of the implicit costs associated with foreign licensing, especially that of managerial time. When time costs were included, and valued at the relevant salary level, they resulted in most early licensing agreements of the Finnish companies being unprofitable.[22]

Australian companies interviewed about the costs of establishing foreign licensing agreements invariably reported that the costs were far greater than anticipated. In addition, there was a range of continuing maintenance costs. The cost and time scale associated with establishing a licensing agreement were related to the need for learning regarding the licensing activity, the demands of locating and selecting suitable licensees, the negotiation process and post-agreement transfers. Overall costs of establishing licensing agreements were, on average, 46.6 per cent of total licensing costs, as against 24.8 per cent for the protection of industrial property and 29.0 per cent for maintenance costs. The main establishment costs were (in order):

communication between the involved parties;
searching for suitable licensees;
training personnel for the licensee.[23]

The costs of achieving an acceptable exchange between the two parties to the licensing arrangement are clearly considerable, reflecting the range of activities which must occur in the initial stages of the transfer process.

Communication requirements

The transfer process in licensing emerges as a highly communication-intensive activity, from initial contacts through to long-run interrelationships between the parties. This is related to the nature of the technology market, which requires a high level of communication if the uncertainty and knowledge gap between potential partners is to be bridged and exchange effected. This is especially the case when the 'commodity' being exchanged, knowledge, is of such an intangible character. By the same token, communication intensity means that there is considerable scope for distortion or disturbance of the communication process, which may interfere with transfer possibilities in various ways. For example,

Australian evidence indicates that the cultural factor is one potentially distorting influence in communication activities.[24] Australian companies tended to feel ill at ease in operating within legal systems of a non-British nature. A number of licensing possibilities were passed up because the country in which the licensee was located was regarded as providing insufficient protection for the firm's industrial property.

The exchange framework

The market for technology is a difficult and uncertain one for buyers and sellers seeking to achieve efficient exchange. The impediments to effective exchange when the parties are unrelated and have no knowledge of each other are considerable. It has, in fact, been suggested that the constraints in the exchange process are supportive of a foreign investment strategy instead.[25] There are major problems at the outset in identifying potential licensees or licensors and obtaining adequate information about their operations, and yet the selection decision is critical to long-run success. If the companies are incompatible and do not work together on the transfer objective, the whole operation is likely to fail. Much depends on preceding experience, of course. In some cases, licensing will follow other forms of foreign operations in the foreign market concerned which reveal a prospective partner, perhaps the company's own representative. To begin the exercise from scratch is more difficult. Australian companies tended to allocate more time and care to the selection process as their experience developed, a recognition of the importance of this step. In interviews it was frequently stressed that selection of a good licensee was far more important than a 'tight' legal document.[26]

At the outset there is also uncertainty about what elements of technical and commercial know-how are saleable, where the appropriate markets are located and what is an appropriate price for the technology being offered. The value of the technology is often difficult for both licensor and licensee to determine.[27] Given the frequently intangible nature of the knowledge asset on offer, the negotiation process can be a demanding exercise before an acceptable agreement is established. A constraint often encountered is that although the potential licensee needs adequate information in order to test the worth and performance of the technology, the licensor is concerned that, having provided this information, the potential licensee will be in a position to use it without entering into an agreement. Secrecy agreements represent only a partial solution to this perceived problem. At all stages preceding agreement a difficult decision must be made by the licensor as to just what level of information needs to be provided to potential licensees. Negotiations may be aborted in situations where excessive concern for secrecy on the part of the licensor limits the ability of the potential licensee to form an effective judgement of the technology on offer. Where the licensor has limited industrial property protection, and the technology has not been commercialised, as is commonly the case with individual inventors, such concern tends to be greater still.

Because of the uncertainties and constraints on the licensor's approach to the technology market, it is not surprising that the initiative for sale often comes from technology purchasers. They may have a clearer understanding of the potential for some new technology in their own market, or they may be seeking a specific technological solution to a problem within the company. In a recent British study it was found that most of the licensing agreements for a sample of small to medium-sized firms were initiated by licensees.[28]

The stages leading up to the signing of the licensing agreement, the formal basis of exchange between the parties, can therefore comprise a difficult, tenuous and time-consuming process, with significant cost implications. In the longer run, success in this licensing activity depends heavily on the acquisition of appropriate skills and knowledge, mainly through experience, which can be applied in successive licensing episodes. Lowe and Crawford concluded that small firms 'without such skills or without access to relevant advice frequently experience substantial problems'.[29]

Organisational fit and interaction

For technology transfer to be effective beyond the signing of the licensing agreement, especially when there is a high degree of intangible know-how involved, a significant level of interaction between the two parties will be necessary. Where there is a high level of disparity between the technological and marketing skills of the licensor and licensee, the transfer and interaction demands are generally accentuated—especially in the case of licensing from advanced to developing countries. However, even between companies in the advanced countries, the further the shift from existing technological and marketing skill areas by the licensee, the greater the amount of learning which will be necessary, requiring extensive personal interaction.[30] As Turnbull has noted, 'personal contacts are at the heart of interaction between organizations and serve as a primary medium of communication in both buying and selling'.[31] In fact, effective technology transfer requires interaction between the parties across a number of dimensions and activities, not just of a technological nature (see Figure 13.1). Social exchange, for example, may be an important interconnected component:

> The need for social exchange is especially significant when the decision makers through social exchange can compensate for a portion of the uncertainty . . . there exists an intimate connection between the physical and social exchange, since the former demands the latter and is also a carrier of same.[32]

Given the interaction demands of the licensing relationship for effecting technology transfer, the fit and compatibility of the parties, the preparedness

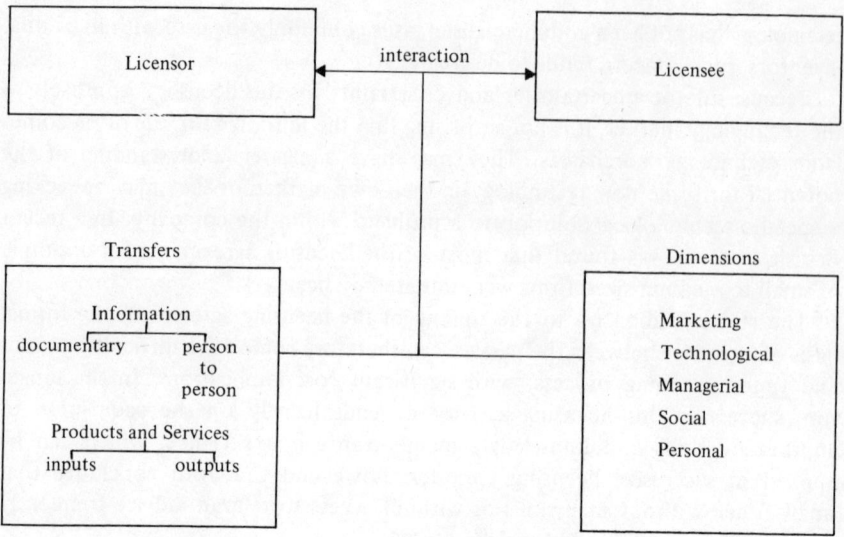

Figure 13.1. An interaction perspective on technology transfer

to commit resources to the relationship by the licensor and the extent of active involvement by both parties in the transfer process are clearly key issues. Technological and market fit has already been noted, but there is a wider perspective of compatibility, covering the total fit of the two organisations. Despite some simple 'rules of thumb' which licensors often employ in determining compatibility with potential licensees, such as the avoidance of competitors in the field, the assessment of this factor is difficult before the relationship is established.[33] It tends to be developed, or not, in the transfer activities of the post-agreement signing phase, and is heavily dependent on the preparedness to react and adapt to the other's requirements in a way which is normally not written down in the agreement itself.

The demands of interaction impose costs on the licensor, which are normally not fully covered by additional payments. These costs impose a limit on the preparedness of the licensor to enter into interaction activities, despite their benefits and the vested interest of the licensor in ensuring successful technology transfer to, and market operations by, the licensee. It is a question of the perceived relative benefit-cost as an ultimate outcome of the relationship. The evidence indicates that licensing is often seen as a marginal exercise, or only used when other more preferred market servicing options are constrained for various reasons.[34] With such a secondary view, it is not surprising that there is sometimes little commitment to the transfer process. In a study of British firms licensing to India, Davies found that:

> ... only a small proportion of the UK collaborators sampled devoted

resources to adaptation. The licensors interviewed were concerned solely with the provision of information on the British product or process, leaving their Indian partner to make its own adjustments or establish a facsimile.[35]

This approach appeared to flow from a view about the peripheral nature of the market concerned, which meant that the returns from greater involvement in and commitment to the transfer process were not considered sufficient to justify the effort.

Thus, the transfer demands of the licensing relationship cause costs which are not always recognised on both sides, nor adequately compensated for in the licensing arrangement from the licensor's viewpoint. Consequently, they remain a significant barrier to the effectiveness of the technology transfer process and help to explain why alternative forms of foreign market involvement may be preferred—by widening the basis of returns they justify a deeper transfer commitment. In general, though, the more compatible the licensing partners, and the more positive and committed both are to the licensing relationship, the more likely it is that the technology will be effectively transferred, implanted and commercialised. A further implication of the stress on compatibility and of the demands of interaction is that licensors or licensees are not so much seeking to buy or sell technology, but are rather seeking a partner with whom to establish an exchange process, based on technology transfer.

Technology transfer in the longer term

Licensing is often viewed as a relationship in which, once the agreed elements have been duly transferred, commitment and involvement cease, apart from general policing of the agreement. Technology transfer, however, is not a single act or episode, but a process. To be effective it normally requires some continuing interaction between the parties. Continued adaptations, adjustments, training and updating of know-how are part of 'technological maintenance'. Of total maintenance costs for a sample of Australian licensors, 65 per cent were concerned with back-up services to the licensee.[36] With such a commitment, the licensing agreement is clearly more attractive for the licensee, while the licensor achieves a measure of positive control over continuing operations, and helps to ensure the long-run success of the licensee on which the ultimate success of the licensing venture depends. In the long run a better understanding of the other's requirements and possibilities tends to grow out of the interaction process, thereby creating a better environment for technology transfer. The learning process is important, not just for general transfer ability, but also within a given relationship.

From a recent study of licensing activity by some Swedish companies, it was concluded that the reason for success or failure had little to do with the licensing object itself, but rather depended on the patience of the parties in building a long-term relationship for mutual benefit.[37] An unsuccessful relationship was seen to be a result of passive involvement. Thus, the effectiveness of technology transfer in licensing has a long-run dimension, which depends on the quality of continuing interaction between the parties. The benefits of continuing interaction are greater than just the assurance of licensing success. This may lead to associated exports, return flows of technology and other valuable information, as well as wider co-operation possibilities. However, the longer term demands and possibility of licensing tend not to be clearly recognised by companies new to licensing, and this constrains its effectiveness as an instrument of technology transfer.

Policy implication

The less developed countries have expressed considerable concern about the terms and conditions of international technology transfer, their concern being reflected in the effort to develop a Code of Conduct on Technology Transfer.[38] Already, though, a number of countries, especially in Latin America, have introduced a variety of controls on technology transfer arrangements in an effort to remove restrictive practices and reduce the costs to the recipients.[39] While the control policies appear to have gone some way towards achieving their objectives:

> more comprehensive policies are necessary if the promotion of technological developments locally is the main objective: indeed for countries which have the capacity and intention to pursue this as a major objective the cost issue —on which so much attention is currently focussed—becomes of subsidiary importance.[40]

The preceding analysis has shown that exchange or relations efficiency between two organisations is an important determinant of the effectiveness of technology transfer. Clearly controls cannot of themselves ensure good relations between the transfer partners. As a result, if policies for technology transfer are to become more effective, they must include a more positive approach towards the building of strong relations between technology buyers and sellers. For example, assistance may need to be provided in the search for, and assessment of, appropriate technology suppliers.

Conclusion

Technology transfer is a highly demanding exchange process, especially when the companies are unrelated and joined only by a licensing arrangement. To be effective, it requires a high degree of commitment by both parties to exchange activities. In general, the greater the interaction, in the short run and long run, the more likely it is that transfer will be successfully accomplished.

However, there are serious constraints on technology transfer in licensing which frequently limit, or even prevent, achievement of the transfer objective.

Licensing is often adopted as a secondary international marketing strategy, and this is reflected in a limited commitment to the activity. More important, though, is the nature of the 'arm's-length' relationship which licensing involves, and the intangible character of the product being exchanged. The distance between the parties in all respects increases uncertainty and constrains contact, negotiation and transfer activities. There are difficulties in defining and valuing the transfer object, and in determining an appropriate price. As many of the demands of effective transfer can never be settled in the agreement, much depends on the way the relationship evolves. Thus, there is considerable room for distortion and misunderstanding in the exchange process, which may seriously interfere with the final outcome. Also all of the stages in effective transfer have considerable cost implications. In many cases, licensors are unprepared to commit adequate resources to the transfer process because the returns are considered to be insufficient to justify the expenditure. Given the trend towards more complex licensing packages, it is likely that the demands and costs of effective transfer will increase.

Notes and references

1. See Helleiner, G. (1975), 'The role of multinational corporations in less developed countries' trade in technology', in Kojima, K. and M. Wionczek (eds), *Technology Transfer in Pacific Economic Development*, Tokyo, Japan Economic Research Centre; Stewart, F. (1979), *International Technology Transfer: Issues and Policy Options*, World Bank Staff Working Paper No. 344, Washington D.C.; Contractor, F. and T. Segafi-Nejad (1981), 'International technology transfer: major issues and policy responses', *Journal of International Business Studies,* **12**: 113-35.
2. See Helleiner, op. cit., pp. 84-5.
3. Contractor, F. (1980), 'The composition of licensing fees and arrangements as a function of economic development of technology recipient nations', *Journal of International Business Studies,* **11**: 47.
4. Stewart, op. cit., p. 19.
5. Australian Bureau of Statistics (1979), *Research and Experimental Development: Private Enterprises, 1976-77*, Canberra, AGPS.
6. Stopford, J. and L. Wells (1972), *Managing the Multinational Enterprise*, London, Longman, pp. 121-2.
7. Teece, D. (1977), 'Technology transfer by multinational firms: the resource cost of transferring technological know-how', *Economic Journal,* **87**: 242-61.
8. Davies, H. (1977), 'Technology transfer through commercial transactions', *Journal of Industrial Economics,* **26**: 161-75.
9. Blumenthal, T. (1976), 'Japan's technological strategy', *Journal of Development Economics,* **3**: 245-55.
10. Stewart, op. cit., pp. 3-4.
11. Oravainen, N. (1979), *Suomalaisten Yritysten Kansaivaliset Lisenssi—Ja Know-How —Sopimukset (International Licensing and Know-How Agreements of Finnish Companies)*, Helsinki, Helsinki School of Economics, FIBO Publication No. 13, p. 35.
12. Oravainen, op. cit., p. 99.
13. Carstairs, R. and L. Welch (1981), *A Study of Outward Foreign Licensing of Technology by Australian Companies*, report prepared for the Licensing Executives Society and Industrial Property Advisory Committee of Australia, revised edition, p. 48.
14. Helleiner, op. cit., p. 82. See also Pengilley, W. (1977), 'Patents and trade practices—competition policies in conflict', *Australian Business Law Review,* **5**: 201.

15. Carstairs and Welch, op. cit., p. 22.
16. Contractor, op. cit., pp. 48-50.
17. Ibid., p. 47.
18. Carstairs and Welch, op. cit., p. 23.
19. Ibid., pp. 22-4.
20. Teece, op. cit., p. 247.
21. Contractor, F. (1981), *International Technology Licensing: Compensation, Costs and Negotiation*, Lexington, Mass., Lexington Books; Oraivanen, op. cit.
22. Oravainen, op. cit., pp. 47-9.
23. Carstairs and Welch, op. cit., pp. 36-9.
24. Ibid., pp. 40-2.
25. Teece, D. (1981), 'The multinational enterprise: market failure and market power considerations', *Sloan Management Review,* **22**: 9-10.
26. Carstairs and Welch, op. cit., p. 30.
27. Killing, J. (1980), 'Technology acquisition: licence agreement or joint venture', *Columbia Journal of World Business*, **15**: 44.
28. Lowe, J. and N. Crawford (1982), *Technology Licensing and the Small/Medium Sized Firm*, Interim Report, School of Management, University of Bath, p. 19.
29. Ibid., Appendix 3, p. 3.
30. Killing, op. cit., p. 39.
31. Turnbull, P. (1979), 'Roles of personal contact in industrial export marketing', *Organisasjon, Marked og Samfunn,* **16** (5): 335.
32. Håkansson, H. and B. Wootz (1979), 'A framework of industrial buying and selling', *Industrial Marketing Management,* **8**: 30-1.
33. Welch, L. and R. Carstairs (1982), 'Some international marketing implications of outward foreign licensing', *Journal of International Marketing,* **1** (3); 177-85.
34. Carstairs, R. and L. Welch (1982), 'Licensing and the internationalisation of smaller companies: some Australian evidence', *Management International Review,* **22** (3): 34-5.
35. Davies, op. cit., p. 173.
36. Carstairs and Welch, op. cit., 1981, p. 38.
37. Wiedersheim-Paul, F. (1982), *Licensing as a Long Run Relation*, Working Paper 1982/2, Centre for International Business Studies, University of Uppsala, Sweden, p. 20.
38. Thompson, D. (1982), 'The UNCTAD code on transfer of technology', *Journal of World Trade Law,* **16**: 311-37.
39. Correa, C. (1981), 'Transfer of technology in Latin America: a decade of control', *Journal of World Trade Law,* **15**: 388-409.
40. Stewart, op. cit., p. 62.

PART IV
FROM EMPLOYMENT TO POLICY

14 Information technology and employment levels
Thomas Mandeville and Stuart Macdonald

Introduction

Microelectronics technology is information technology in that it stores and processes information. Some applications, such as word processors, are much more clearly seen to be handling information, but the chip controlling a modern sewing or washing machine is really doing exactly the same thing. Since the invention of the transistor in 1947, the technology for making thousands of almost simultaneous switching decisions has developed and diffused at a furious pace.[1] Neither development nor diffusion shows any sign of slackening and, not surprisingly, great concern has been expressed about the impact of so much microelectronics technology on employment levels. Rising, rather than high, unemployment rates have exacerbated this concern. Especially significant, however, has been the growing recognition of the importance of what it is now acceptable to call the information sector of the economy: information technology seems to be a very direct threat to employment in the information sector, a sector which has been expanding rapidly and which now employs about a third of the workforce in many developed countries.[2] In particular, the conjunction of information technology with telecommunications—itself using the same microelectronics technology—much reduces the constraints of location on the impact of information technology.[3] The result has been the achievement of sophisticated technical ability to handle information, unmatched by equal advance in either appreciation of the importance and characteristics of information as a good, or understanding of the process of technological change. It has been common to perceive technological change as a discrete entity, a hardware package from the R & D laboratory. In fact, it is much more realistic, though less neat, to see technological change as simply the adoption of what seems to be a better way of doing things, or what seems to be a good way of doing new things altogether. Such a definition would probably not have displeased Schumpeter,[4] though there is now emphasis on perceived, rather than actual, improvement. The adoption of an innovation to deal with an unfamiliar good which is also inherently uncertain is particularly likely to be determined by perceived benefits rather than real benefits. In such a confusing milieu, it can be a little difficult isolating the links between information technology and levels of employment.

Perceptions of the problem

Paradoxically, most observers have experienced no trouble at all in identifying the impact of information technology on employment. Most people take the view that information technology simply results in the direct displacement of employees in the innovating firm or industry; machines enter and people leave, through either outright redundancy or natural wastage. This view is amplified by films such as the seminal British television documentary *Now the Chips are Down*, by dramatic assertions in the media,[5] and case-studies of certain information industries, especially telecommunications and the newspaper and printing industries.[6] The orthodox notions of orthodox economists have done little to calm passions.[7] Most economists take the view that technological change must increase productivity, that it would hardly be introduced if this were not the case. Technological change, they reason, may well result in the displacement of a few workers in the innovating organisation, but it also produces higher profits or reduced production costs and, therefore, at least potentially, reduced prices. Either way, there should be an increase in real income which should be reflected in greater demand for goods and services. In order to provide these extra goods and services, so the argument goes, more labour must be recruited and this should compensate for the initial labour displacement occasioned by the technological change.

Of course, the general experience since the Industrial Revolution has most certainly been that technological change is often productivity-enhancing and labour-saving in the innovating firm or industry. Agriculture in the developed world provides perhaps the best sectoral example of this.[8] The crucial question here, though, is whether the same happy pattern will be repeated with microelectronics technology in an emerging information age. Increased food production through gradual agricultural mechanisation may not provide a universally applicable model. What if the production of even new goods and services requires little additional labour, or little of the labour of some countries rather than others? What if real incomes are higher for a few, and actually lower for many? What if reduced costs are not transformed into lower prices and, indeed, what if new information technology does not reduce total costs at all? The last point is particularly important and we will return to it later in this chapter.

Orthodox economics, however, continues to provide the theoretical structure for government inquiries into the act of technological change, and for ill-advised and ill-founded attempts to model its economy-wide effects.[9] In spite of the importance of the issue, remarkably little effort has gone into what Liebenstein advises that economists should do, namely going out and looking.[10] We have been fortunate in having the opportunity over the last few years to do a good deal of going out and looking at information technologies and employment— in banking and insurance, local government, the technology supply industry, as well as in small business.[11]

Technological change may be rapid, but social and organisational adjustments tend to be much slower.[12] At the time of adoption, the ultimate outcome is always uncertain. The actual impact of information technologies is usually quite different from that envisaged by their most enthusiastic promoters or detractors. Our own studies illustrate this point; they suggest that the adoption of new information technology commonly starts slowly with the new technology performing just a few existing tasks, particularly those imposing a strain on the system and to which the diversion of extra resources would bring only diminishing returns. This should not be seen as the end of the process. Technological change is, above all else, a dynamic process: change begets more change. As innovators become familiar with the strengths and weaknesses of the new technology, they extend its use to further functions and sometimes to new ones altogether. With experience, and particularly experience of specific market demand, the technology improves and becomes more attractive and useful to the market. Other technologies develop altogether and it is often the synergistic relationship of two technologies—computers and telecommunications are a good example—that results in radical innovation. There is no basis for any assumption that all this will necessarily happen at a slow, measured and predictable pace in a predetermined direction, or even that the innovators themselves are always in a position to know what the pace and direction will be.

The situation becomes even more confusing and uncertain when there are rapid improvements in technologies and when those technologies are readily finding new markets in all sectors of the economy virtually simultaneously. This would appear to be the case with much modern information technology.[13] Furthermore, there are many factors other than technological change which influence economic activity and employment; for example, the state of the world economy, aggregate demand, inflation, educational capacity, and labour relations. In these circumstances, no one should pretend to know with any certainty what will be the eventual impact of information technology on employment. At best, only tentative suggestions can be made.

Direct employment effects

Our own work suggests there is little direct displacement of labour as a result of the introduction of information technology.[14] For example, our study of the word processor supply industry in Australia finds a very rapid diffusion of that technology and yet little reduction in the total numbers of typists and stenographers.[15] That is not to say that these employees will continue to perform traditional tasks, but their new functions will not be predetermined by the technology. Rather they will depend on the use to which the innovating organisation puts the word processor. Similarly, our studies of computer usage in small business and in local government discover little evidence of direct job losses as a consequence of computer adoption.[16] In the banking industry, we find that new technology has helped the banks diversify further into areas that are not traditional to banking.[17] Overall staff numbers have been maintained, but a survey of bank branches in Queensland suggests that 20 per cent of staff time is now spent on such new activities as travel, insurance and dealing with credit cards. In a sense, 20 per cent of bank employees no longer work in the banking industry, at least not the banking industry as it was before new information technology facilitated the expansion of these activities. Such technological change demands more flexibility from both employees and the educational system; it demands a readjustment of career expectations; it brings disappointment to some employees and opportunities to others, but above all else it brings uncertainty.[18]

Whenever specific information technology—such as a word processor, automatic telling machine, robot, or even a computer—becomes the focus of debate, the notion of 'displacement ratios', of how many employees each machine will directly replace, is often evident. Of course, the concept of displacement ratios follows logically from the widespread assumption that information technology directly substitutes for people. In Australia, Thornton and Stanley have used an extraordinary calculation to reveal that the computers already installed in the country by 1978 were doing the work of between three million and twenty-one million people. At the same time they argued that: 'to be worthwhile from an economic investment point of view, the huge investment so far . . . would have had to have been able to displace some 200,000 jobs to have been worth doing at all.'[19] The argument that it is worthwhile installing a computer only when its cost is exceeded by the salaries of those displaced is, of course, quite untenable when linked with the argument that installed computers do more work than the people they displace. If computers are responsible for so very much more productivity, one wonders why a firm is forced to displace existing staff to justify computer installation.

It is naive and generally misleading to talk of manpower equivalents of computers. The horsepower equivalent of engines has long since ceased to have anything to do with the number of horses displaced. Computers and people are much more complements than substitutes. Unaided humans cannot feasibly perform thousands of calculations in seconds. On the other hand, computers comprehend only exact instructions and thus have great trouble with aspects of reality difficult to programme—such as subtle shades of meaning, heuristics and intuition, pattern recognition and the ability to screen out, sensitively, subtly and intuitively inessential information in order to arrive at the heart of the matter.[20] Leibenstein, the architect of X-efficiency theory, casts sober light on the mechanical application of economy-wide displacement formulas to new technologies:

> you cannot tell how much more productive new equipment will be because that depends on the effectiveness of the employees' effort choices with the

new equipment compared to the old. Thus the usual formulas cannot readily be applied.[21]

One of the few serious, comprehensive, and survey-based studies of the employment displacement effects of mainframe computers was carried out by the UK Department of Employment in 1972.[22] Its findings were that net displacement effects had been trivial—125,000 persons out of the total British labour force of 25 million.

Word processors apply computing techniques to office typing. Those who fear that word processors are causing unemployment typically claim that each machine installed will directly displace between two and five secretaries.[23] But a diversity of modern word-processing equipment is available, and its performance is dependent on such other factors as type of industry, office organisation and associated technologies. Detailed assessment of such factors is essential before it is possible to say anything very precise about productivity increases resulting from the use of word processors. Much of a secretary's, or even a typist's time, is occupied by a wide variety of miscellaneous tasks apart from typing. Three basic job functions of typists/secretaries are word transmitting, acting as a go-between, and providing back-up services. Displacement ratios, if they are a valid concept at all, are likely to apply only to the word-transmitting function. Even here, though, the experience of users suggests that word processors may be work-generating.[24] Following Parkinson, typing work in the office environment may expand to fill the time or resources available. Word processing allows the opportunity to revise material or to originate new material where this, previously, was hardly feasible. While established methods have been tailored over time to tackle traditional tasks, that cannot be the case with an innovation. Information technology introduced to cope better with traditional tasks usually makes possible the performance of entirely new tasks; indeed, it often suggests or even creates them.[25]

Our study of technological change in the insurance industry supports the proposition that direct employment effects can depend on the way firms choose to use the technology.[26] Insurance firms associate technological change with reduced staff numbers, and increased staff numbers with a growth in demand for insurance services. There is apparently little awareness of the importance of technological change in allowing satisfaction of this demand. It is conceivable that insurance companies could exploit their technological capability to offer new goods and services to established markets—as the banking industry has done—and to penetrate new markets with both these and the industry's traditional products. However, there is little evidence that the Australian insurance industry is using its considerable technological base for anything other than the provision of traditional goods and services to traditional markets. Under these circumstances, technological innovation is indeed likely to be labour-saving, and employment levels in this industry are very likely to fall.

Indirect employment effects

Our study of computers in small business provides some indication of important indirect effects on employment levels resulting from failure to innovate. A survey of computer use in nearly 2,000 small businesses found that firms using computers had increased their employment over the 1974-1979 period.[27] Employment in firms which had failed to adopt computers had declined. It is possible that those firms which become more competitive through innovation grow at the expense of the less competitive laggards. If that is the case, then technological change may indeed be displacing employees—but in non-innovating, rather than innovating, firms.

Small businesses using computers were also asked what effect they thought their new technology would have on their future employment levels. Over 80 per cent imagined that, as a direct consequence of their use of computers, they would experience no growth in employment. The phenomenon by which new technology allows production to increase without concomitant increase in employment is coming to be known as 'jobless growth'.[28] There is nothing very clever or novel in the notion, and economies normally thrive on such a process. But when the process is occurring rapidly and across many sectors of the economy simultaneously, and when there is increased production of new goods and services for which there is no established market, jobless growth may well be accompanied by serious employment consequences.

However, the phenomenon of jobless growth depends upon organisations using information technologies efficiently to increase productivity. There are good reasons for doubting that this will inevitably occur as more and more firms struggle to utilise the massive and often unanticipated capacity of information technologies. It may be no coincidence that the widespread diffusion of intelligent electronics has coincided with a significant reversal of long-term trends in productivity gains. Soete refers to the paradox of much of the employment and technological change debate: the widespread diffusion of supposedly productivity-enhancing electronics technology in the midst of low and falling overall productivity growth.[29] Denison estimates that the contribution to US productivity of 'advances in knowledge and miscellaneous determinants' (the residual encompassing technological change) has been negative since 1973.[30] While providing a catalogue of possible explanations, including the oil crisis of that year, he is forced to confess that he does not know why there should have been such a marked change for the worse.

It is possible to provide a tentative explanation of how it is that productivity can decline in the face of the rapid adoption of productivity-increasing technology, an explanation which considers the limitations of information technology and which links several of Denison's major causal factors. Through what we have called the 'xerox effect', information technologies may be eroding overall information efficiency.[31] New information technology is adopted by the organisation to perform specific tasks; however, the enormous capacity potential inherent in these technologies enables them to perform quite a number of other tasks. The more tasks the technology performs, the more strain is placed on the existing structure of the organisation and the greater is the necessity for the organisation to adjust and adapt—otherwise information will be produced indiscriminately and will be used inefficiently. Changes in organisation structure to make use of newly available information may be difficult and expensive.[32] However, if the organisation fails to adapt, the efficiency with which it handles all its information may be impeded, with consequent reduction in productivity.

One can envisage a vicious circle in organisations attempting to cope with a flood of internally generated information by hiring more knowledge workers —accountants, forecasters, analysts, programmers, consultants, researchers, administrators and co-ordinators—who in turn generate more information and demand more support from information technology.[33] Normally, of course, the market mechanism would weed out organisations bogged down in self-generated work—unless, of course, the cause of the work is a prerequisite for remaining in the market and competitors are experiencing similar novel difficulties.[34] This may well be the situation. Indeed, Strassman, Vice-President of Xerox Information Products Group, has argued that:

> Transformation from the office of today to a more efficient office of the future seems elusive, and the hypothesized result is by no means assured. Movement towards our new goal turns out to be more difficult and complex than the changes that were necessary for industrialization.[35]

If this 'xerox effect' hypothesis approximates the true situation, then concern about the effect of information technologies on employment levels is hardly likely to reach the heart of the problem. Indeed, the problem becomes not the utopian one of how to distribute excess wealth, but the old, familiar one of how to create it. This is likely to involve the creation of new types of organisation, as well as new forms of measurement, so that output fully reflects increased information input. Employment levels are dependent on ability to make economic use of information. Contrary to popular belief, new information technology does not compensate for inability to use information efficiently. In fact, it seems to aggravate that inability, and that situation—especially if its existence is scarcely recognised—has very serious implications indeed for employment levels.

Notes and references

1. Freeman, C. (1979), 'Social and economic impact of micro-electronics', paper to the Technology Assessment Workshop, Department of Science and the Environment, Sydney.
2. OECD (1981), *Information Activities, Electronics and Telecommunications Technologies*, Paris, Vol. 1.

3. Mandeville, T. (1983), 'Spatial effects of information technology', *Futures*, 15 (1): 65-72.
4. This definition bears some resemblance to Schumpeter's notion that technological change is 'any "doing things differently" in the realm of economic life'. See Schumpeter, J. (1939), *Business Cycles*, New York, McGraw-Hill, Vol. 1, p. 84.
5. Australian examples are critically discussed in Mandeville, T. and S. Macdonald (1980), 'Reflections on the technological change debate in Australia', *Australian Quarterly*, 52 (2): 213-20. A useful appraisal of some of the literature occurs in Braun, E. and P. Senker (1982), *New Technology and Employment*, London, Manpower Services Commission; and Ford, W., M. Coffey and D. Dunphy (1981), *Technology and the Workforce*, Sydney, Technology Research Unit.
6. For example, Rothwell, R. and W. Zegveld (1979), *Technical Change and Employment*, London, Frances Pinter; Forward, P. and G. McColl (1980), 'Technological change in the Australian newspaper industry' in *Technological Change in Australia (Myers Report)*, Report of the Committee of Inquiry into Technological Change in Australia, Canberra, AGPS, Vol. 4, pp. 143-88; Bennett, E. (1979), *New Technology and the Australian Printing Industry*, Sydney, Printing and Kindred Trades Union.
7. See, for example, Australian Treasury (1979), *Technology, Growth and Jobs*, Economic Paper No. 7, Canberra, AGPS.
8. Rosegger, G. (1980), *The Economics of Production and Innovation*, New York, Pergamon Press, pp. 299-300.
9. For a critique of one such attempt, see Mandeville, T., S. Macdonald and D. Lamberton (1981), 'The fortune-teller's new clothes: a critical appraisal of IMPACT's technological change projections to 1990-91', *Search*, 11 (1/2): 14-17.
10. Leibenstein, H. (1982), 'The state of economics', seminar paper presented to the Department of Economics, University of Queensland.
11. Macdonald, S., D. Lamberton and B. Hodge (1981), *Tradition in Transition—Technological Change and Employment in Banking*, Working Paper No. 33, Department of Economics, University of Queensland; Brown, A. and S. Macdonald (1982), 'Technological change and employment in the insurance industry', *Economic Activity*, 25 (4): 6-20; Mandeville, T. and S. Macdonald (1981), 'Computers and employment in local government' in *Preprints of Papers of the First National Local Government Engineering Conference*, Adelaide, pp. 105-10; Macdonald, S. and T. Mandeville (1980), *Diffusion and Employment Effects of Word Processors in Australia*, Working Paper No. 21, Department of Economics, University of Queensland; Macdonald, S. and T. Mandeville (1980), 'Word processors and employment', *Journal of Industrial Relations*, 22 (2): 137-48; Macdonald, S., T. Mandeville and D. Lamberton (1980), *Computers in Small Business in Australia*, Industry Economics Discussion Paper No. 14, Department of Economics and Institute of Industrial Economics, University of Newcastle.
12. See Eliasson, G. (1982), 'Electronics, economic growth and employment—revolution or evolution?' in Giersch, H. (ed.), *Emerging Technologies: Consequences for Economic Growth, Structural Change and Employment*, Tübingen, J. Mohr, pp. 77-95.
13. Stout, D. (1980), 'The impact of technology on economic growth in the 1980s', *Daedalus*, 109 (1), 159-67; Barron, I. and R. Curnow (1979), *The Future with Microelectronics*, London, Frances Pinter.
14. This finding is supported by other studies, for example, Manpower Research and Information Branch, Department of Employment and Youth Affairs, New South Wales (1979), *Word Processing and Some Aspects of Its Employment Impact on the Typing/Secretarial Area*; Department of Labour and Immigration (1975), *Studies of Displacement*, Employment and Technology Series No. 16, Canberra, AGPS. See also several other publications in the same series, especially the *National Survey of the Employment Effects of Technological Change*, Stages 1-5.
15. Macdonald, S. and T. Mandeville (1980), *Diffusion and Employment Effects of Word Processors in Australia*, Working Paper No. 21, Department of Economics, University of Queensland.
16. Macdonald, S., T. Mandeville and D. Lamberton (1980), *Computers in Small Business in Australia*, Industry Economics Discussion Paper No. 14, Department of Economics

and Institute of Industrial Economics, University of Newcastle; Mandeville, T. and S. Macdonald (1981), 'Computers and employment in local government' in *Preprints of Papers of the First National Local Government Engineering Conference*, Adelaide, pp. 105-10.
17. Macdonald, S., D. Lamberton and B. Hodge (1981), *Tradition in Transition—Technological Change and Employment in Banking*, Working Paper No. 33, Department of Economics, University of Queensland.
18. See Evans, O. (1980), 'Implementing technological change', *Australian Journal of Public Administration*, **39** (2): 221-9.
19. Thornton, B. and P. Stanley (1978), *Computers in Australia—Usage and Effects*, Sydney, Foundation for Australian Resources, p. 11.
20. See Weizenbaum, J. (1976), *Computer Power and Human Reason*, San Francisco, W. H. Freeman and Co.; Marschak, J. (1971), 'Economics of inquiring, communicating, deciding' in Lamberton, D. (ed.), *Economics of Information and Knowledge*, Harmondsworth, Penguin, pp. 37-58; Macdonald, S. (1980), 'The impact of technological change', paper delivered to the Council of Social Service of Ipswich Public Seminar.
21. Leibenstein, H. (1979), 'X-efficiency: from concept to theory', *Challenge*, **22** (4): 19.
22. UK Department of Employment (1972), *Computers in Offices 1972*, Manpower Studies No. 12, London, HMSO.
23. For example, Manpower Research and Information Branch, op. cit.; Thornton and Stanley, op. cit.; Kornhauser, C. (1978), 'Bid to control introduction of word processors', *Australian Financial Review*, 13 December: 17.
24. Macdonald and Mandeville (1980), *Diffusion and Employment Effects of Word Processors in Australia*, Working Paper No. 21, Department of Employment, University of Queensland.
25. See Central Policy Review Staff (1978), *Social and Employment Implications of Microelectronics*, London, pp. 7-8.
26. Brown, A. and S. Macdonald (1982), 'Technological change and employment in the insurance industry', *Economic Activity*, **25** (4): 6-20.
27. Macdonald, S., T. Mandeville and D. Lamberton (1980), *Computers in Small Business in Australia*, Industry Economics Discussion Paper No. 14, Department of Economics and Institute of Industrial Economics, University of Newcastle.
28. Freeman, C. (1977), 'Technical change and employment', paper presented to the Conference on Science, Technology and Public Policy: An International Perspective, University of New South Wales.
29. Soete, K. (1981). 'Technical change, international competition and employment', paper prepared for the joint conference on Technological Industrial Policy in China and Europe, Research Policy Institute, Lund, Sweden.
30. Denison, E. F. (1979), 'Explanations of declining productivity growth', *Survey of Current Business*, **59** (8): 1-24.
31. See Lamberton, D., S. Macdonald and T. Mandeville (1982), 'Productivity and technological change: towards an alternative to the Myers' hypothesis', *Camberra Bulletin of Public Administration*, **9** (2): 23-30. See also Soete, op. cit., p. 3.
32. See Arrow, K. (1974), *The Limits of Organization*, New York, Norton, p. 49; Arrow, K. (1979), 'The economics of information' in Dertouzos, M. and J. Moses (eds), *The Computer Age: A Twenty Year View*, Cambridge, Mass., Harvard University Press, p. 315.
33. Porat has observed that knowledge workers are not terribly productive. See Porat, M. (1977), *The Information Economy*, Washington D.C., US Department of Commerce, Vol. I, p. 183; see also MacLaughlin, R. (1978), 'The (mis)use of DP in government agencies', *Datamation*, **24** (7): 147-57.
34. See Withington, F. (1980), 'Coping with computer proliferation', *Harvard Business Review*, **58** (3): 152-64.
35. Strassman, P. (1980), 'The office of the future—information technology for the New Age', *Technology Review*, **82** (3): 55.

15. Trade unions and technological change
John Corina

Trade unions in the 1980s, as in the past, display neither a simple nor a stable response to technological change processes. The complexity of their reactions, shaped partly by bitter experiences of changing frontiers of industrial conflict and partly by innate caution, may be judiciously interpreted as a traditional mode of pragmatism, since trade-union opposition to technology *per se* (or uncritical disregard of likely paths and consequences of technological change) would scarcely comprise policy stances consistent with traditional trade-union goals of preserving jobs and raising living standards.

Unions, technology and the workplace

Yet there is also a more theoretical justification for the complexity of response. National union pronouncements on technology impacts are no guarantee of appropriate rank-and-file orientation and behaviour at the workplace. Worker responses to technology have long been identified as comprising a complex interaction pattern (the 'sociotechnical system') which reflects the nature of the organisation as a combination of technology (equipment, physical layout, task requirements) and of a social system (systematic relationships between those who perform the tasks). The technology and the social system are in mutual interaction, and each is to some degree a determinant of the other.[1] The thrust of much empirical research has therefore been focused upon expanding and refining definitions of technology, and on identifying intervening variables between technology, attitudes and behaviour.[2] Any collective response to technology is now analytically seen as the part-product of workgroup norms, expectations, history and constraints: all factors bearing upon formal and informal organisation at the workplace. As a concomitant, it has long been recognised that the rhetoric of official trade-union responses to technology seldom corresponds with many realities of workplace behaviour.[3] Nevertheless, the OECD (at the end of a decade in which unemployment rates doubled) has not been slow to warn trade unions that: 'Luddite policies or attitudes of resistance to technical change would be counterproductive, as all the problems of generating new employment opportunities would be exacerbated by a slow rate of growth or failure to maintain competitive performance.'[4] However, the industrial relations problem dimensions of technological change are manifestly different in character from general problems of economic management. Most of the evidence suggests that it is the power impacts of advanced technologies upon the varied processes of establishing and administering the 'web of rules' of the workplace which will tend to form the key problem areas in industrial relations systems during the 1980s.[5]

The trade-union response

Initial responses of unions in industrialised market economies to the newer questions of microelectronic technology and employment have been markedly imprecise for further reasons. Policy formulation has been impeded by poor data resources, information deprivation, and conditions of uncertainty severe enough to render forecasting a highly unreliable foundation on which to base policy proposals of any specificity. This deficiency is not so much an intrinsic attribute of the union movement's shortcomings as a deficiency of the society within which the unions function. Unions have found themselves in the centre of a controversy over the employment displacement effects of new technology, where conflicting claims (unable to predict the net employment outcomes of technological change) have polarised interpreters into two broad categories, supporting either the negative (job destruction) or positive (job creation) effects of new technology.[6] Labour spokesmen have a strong tendency to stress the negative aspects while management spokesmen stress the positive inclination. Although it is perhaps unwise to suppose that finely accurate quantitative predictions can be invoked upon the intersectoral employment impact of the new technology for individual national economies, what may be possible is for unions to press for a larger data base, upon an agreed methodology for roughly estimating sectoral impacts, and for monitoring of employment situations for warning signals. It would then be possible for the more sophisticated union centres to construct scenarios of the impact of, for example, microelectronics on employment (ranging from optimistic to pessimistic), each based on explicit assumptions about the diffusion rate of microelectronics, international technology transfer, capital requirements, the social adaptation rate and other sensitive variables.[7]

Generally, trade-union policy statements and reports have highlighted the 'job-destroying' implications of technology, and have presented proposals to modify either technology policy or economic policy, or both, to alleviate what unions perceive as a major unemployment problem gaining in intensity.[8] It is a useful, though second-best role, if only to counterbalance the unwarranted optimism often displayed by governments and employers. In Europe, formal policy statements have been produced, for example, in the United Kingdom by the TUC,[9] as well as by a number of individual trade unions such as APEX,[10] the ASTMS,[11] BIFU,[12] and the POEU.[13] A major French report was produced in

1977 by the CDFT,[14] while in Scandinavia (as well as individual statements by national centres) a special working group of the Nordic Council of Trade Unions has been set up to monitor developments in technology.[15] Trade unions in most other European countries (for example, Austria, the Federal Republic of Germany, Italy, Belgium and Switzerland) have produced research and educational material analysing the impact of technology in their national economies, or providing policy guidelines for negotiating officers. At the European level, a major report has been issued by the European Trade Union Institute, which has markedly influenced subsequent collective bargaining trends.[16]

Outside Europe, interest in technological change has been greatest in Australia, New Zealand and Japan. Virtually every major trade union has produced policy material and union representations in Australia (especially in telecommunications) led to the establishment of a Committee of Inquiry (Myers Committee) into the process of technological change.[17] Partly in response to the energetic approach of European trade unions, American unions have also given serious thought to technological change. Growing understanding that technological questions cannot be evaluated in a fragmented or isolationist fashion has introduced a flurry of international activity among unions. Quite apart from the European level, international trade union secretariats including FIET,[18] the International Metalworkers' Federations (IMF),[19] and the Postal, Telegraph and Telephone International (PTTI)[20] have issued notable studies and policy statements or held conferences of national union specialists. Other international union bodies such as the International Confederation of Free Trade Unions (ICFTU), the Trade Union Advisory Committee to the OECD and regional bodies of the ILO have also made pronouncements.

A composite 'trade-union view' has thus emerged, largely from diverse national union structures facing common fundamental questions, where technological change is usually symbolised by microelectronics and robotics.[21] Is the rate and diffusion of technological change increasing? Is new technology fundamentally different from that of the past? Can the institutions of a market economy cope with change while ensuring a return to full employment and more efficient resource allocation? How are the costs and benefits to be distributed, especially between employed and unemployed, and between private and social claims on GDP? Who controls the new technology? How will new technology affect hierarchy and work organisations? The underlying union social values may be summarised as suggesting that there should be human choice in the introduction and use of technology, where technology is seen not as an external variable but as part of the social system, neutral in the sense that it is to some extent amenable to political and economic influences.

Some common prescriptive components comprising the 'trade-union view' may be discerned as follows:[22]

(a) A policy of blanket opposition to technological change is neither practical nor desirable. No economy can maintain employment by ossifying technology and hampering the productivity growth which provides scope for rises in workers' living standards. Unions seek the establishment of rights to participate in the change process through effective joint consultation (as in the case of the German *Betriesrat*), collective bargaining (as in the case of the Swedish co-determination law), or arbitration provisions (as in the case of Section 88G of the New South Wales Industrial Arbitration Act).

(b) Unions firmly believe that new forms of microelectronics technology in the factory and office are likely, unless forestalled by positive measures, to induce a loss of jobs in the short run. Opinions vary about the extent of job reduction, the time scale and the sectors most affected, but there is some consensus that 'flexible manufacturing systems' will severely affect medium- and small-scale industry, and that white collar and service workers are at high risk from the new microelectronics.

(c) The major responsibility for avoiding large-scale unemployment is seen to be governmental. With rising unemployment, the unions have not predominantly blamed technological change as the key factor. Rather they focused on economic constraints and policy deficiencies that have stopped the process whereby technological innovation may be converted into real income growth rather than unemployment. However, unions have still to distinguish between the concept of preserving jobs (the particular set of tasks performed by a worker) and the more important concept of preserving employment. Unions generally place a high premium on employment security, but job security guarantees vary widely amongst countries. Most Australian employees, unlike British employees, work under awards which give no legal entitlement to any special treatment in a redundancy situation. This is a major reason why technological change is often a fertile source of industrial conflict in Australia.[23]

(d) Unions have not yet reassessed fundamental attitudes to work in the light of released resources from the new technologies. The so-called 'collapse of work' thesis remains a remote horizon.[24] Nevertheless, there is a widespread acceptance that the shrinking volume of available man-hours worktime should be translated into shorter working time for more people. Some unions describe this as 'work-sharing'; others object to the term. Reductions in the working week, longer holidays, flexitime and earlier retirement are all traditional trade union objectives.

(e) Unions generally agree that governments should attempt to brake and discourage the labour-displacing effects of technology.[25] Some unions (as in Sweden) favour a technology tax to adjust the relative costs of capital and labour. The producers of new technology, as well as user organisations, should become more aware of social responsibilities. Job security agreements should be negotiated with employers. The capacity 'to avoid or minimise social costs may prove just as beneficial as realisation of direct productivity gains'.[26]

(f) A consensus exists among unions, regardless of national origins, that workers should be protected against de-skilling, and against technology which could dehumanise work (for instance, by subjecting workers to electronic control and surveillance) or adversely affect health and safety. Misgivings concerning health effects are not without substance. The hazards of work-related stress, already a widespread problem, may increase with a rise in the number of stressors in the work environment. The World Health Organisation warned in 1982 that visual display units (VDUs) can have severe adverse effects on the health of operators.

(g) One key aspect consistently articulated by trade unions is the proposal that workers should possess the right to participate in decisions about the application of new technology in the workplace. The general Australian practice is for little or no consultation with employee representatives prior to the introduction of technological change, and few firms appear to be aware of the National Labour Consultative Council's guidelines on notification and consultation. Some unions go so far as to suggest specific legislation giving the unions or workers an effective veto. The Australian Council of Trade Unions' proposal of a five-year moratorium on all job-displacing technology can be considered as an extreme example of this approach. There are many ways in which a new technology system can be introduced, and where managers present workers with a *fait accompli*, this reflects economic, social and political decisions which have been cloaked in the guise of technology to sustain managerial prerogatives and pre-empt early consultation.

(h) Lastly, most unions concur that union representatives should be afforded training in the basic ideas of computer and telecommunications technology. Systems design should become a joint process, with union representatives and management setting some of the parameters for technologists. The community at large should be made more aware of the social implications of computerisation and the new materials technologies.

Strategy for new technology?

The British TUC provides a leading illustration of union strategy in the light of the awareness programmes developed in Europe. At the shop-floor negotiating level, the TUC is attempting to protect the jobs, job content and incomes of affiliated union members through collective bargaining.[27] This presupposes that most technological change is negotiated and is not the exclusive province of managerial prerogative. It assumes information on corporate planning as well as the development of workplace negotiating machinery. The specific response of individual unions has been to adopt model agreements setting out the terms which negotiators should secure in major areas.[28] But the negotiated concessions in the new technology agreements, reached so far, have been far distant from the goals of model agreements. Unions have not generally secured an associated reduction in hours. They have seldom won an increase in earnings for using new technology, although agreements usually ensure no reduction in earnings and no downgrading for those offered redeployment. On the other hand, unions have generally obtained guarantees that compulsory redundancies will not follow new technology. Some unions have also negotiated substantial concessions: that work performance will not be monitored by computer systems, that work subcontracting should not be introduced, that shiftwork will not be unilaterally introduced, that jobs will not be de-skilled, that training and attendance payments will be provided, and that there should be membership agreements. However, most British unions have been unable to achieve such situational concessions, partly a reflection of differences in bargaining power and labour market conditions during a major recession. There can, therefore, be no presumption that new technology will bring a steady stream of positive benefits to union members via the fashionable negotiating of new technology agreements. This assumption of strategy reflects a reactive approach, characteristic of unionism, of opting for adaptation in the expectation that membership will experience minimum losses. However, since union strategies positively concentrate upon the job security of the job-holder, they consciously require wider social mobilisation techniques to preserve and expand employment, perhaps even implying the radical restructuring of government policies. It is thus simplistic to present unions as intrinsic opponents of change when so often they are mere reactors to uncertainty.

While unionism seeks to control and soften the impact of new technology, in turn, the challenge of new technology has itself become a new factor, influencing the organisation, strategy and goals of unions. Two trends are discernible. First, many British unions are undertaking intensive reviews of their organisation and structure in the light of projected falls in membership arising from unemployment and job-displacement effects of new technology. Servicing members on a shrinking financial base has tended to alter the established distribution of branch and regional power within certain unions. Within many unions, there has also been a marked increase in the influence of National Technology Committees upon Executive Council policies, a redefinition of the significance of plant and company bargaining representatives, and a renewed interest in inter-union co-operation on the international plane.[29] Nevertheless, in some national economies, such as Australia, unions remain unduly fragmented in terms of size and membership composition, although technological change has emphasised the need for amalgamation and the strengthening of central union bodies.[30]

The second effect of the new technology wave has been to induce a higher degree of sophistication in union bargaining techniques (not always matched by employers), best illustrated in the British APEX strategy upon the introduction of word processing and other office automation. The diagrammatic guide (Figure 15.1), now frequently consulted to assist union representatives through the complex decision system, is based on the assumption that:

> Word processing provides the potential to substantially reduce the number of manhours required to do a given amount of work in the office. It could mean a better, more skilled, more satisfying, shorter and less arduous working life for the office worker. It need not lead to unemployment and indeed it should not.[31]

The negotiating of technology agreements through supplementary collective bargaining procedures has become a specialised function, producing in some European trade unions the phenomenon of the 'technology shop steward' (especially in Norway). Elsewhere, unions have evolved extensive resource pools of representatives experienced in presenting employee views upon consultative company-level bodies such as the 'technology conference' and the 'observation review'—organs which were non-existent a decade or so ago.

The role of labour unions

The trade unions' problem is that they have so far construed their goals in a world where full employment and high labour force participation rates have been dominant, and now they are confronted with two constituencies whose interests may conflict—the employed trying to preserve living standards, and the unemployed seeking work. One scenario is the silicon dream of a *passé* work ethic, increased leisure and higher material consumption. Another is of a working population triply split: between those with and those without work, the latter experiencing estrangement, and the former comprising both a group flourishing as the aristocrats of the new technology, and also a group experiencing more intense work uncompensated by requisite increases in real earnings or falls in hours. Whatever the unfolding outcome, unions during the transition phases are likely to adjust their balance of emphasis away from the defensive role of protecting pay and conditions towards the political-interventionist role of pressing for more equitable shares in the resources of industrial and community power.

Issues concerning the quality of working life are not always easy to unravel from the power elements involved in the social control of work organisation. The evolution of information technologies affects both aspects. The new information technology presents a fresh set of choices to unions since it makes possible more decentralised organisations. However, subunits of the organisation are not completely autonomous, but interdependent throughout the network. Thus, the precise organisation form that matches the technology may not be fully determined. It could be centralised or decentralised, but the social

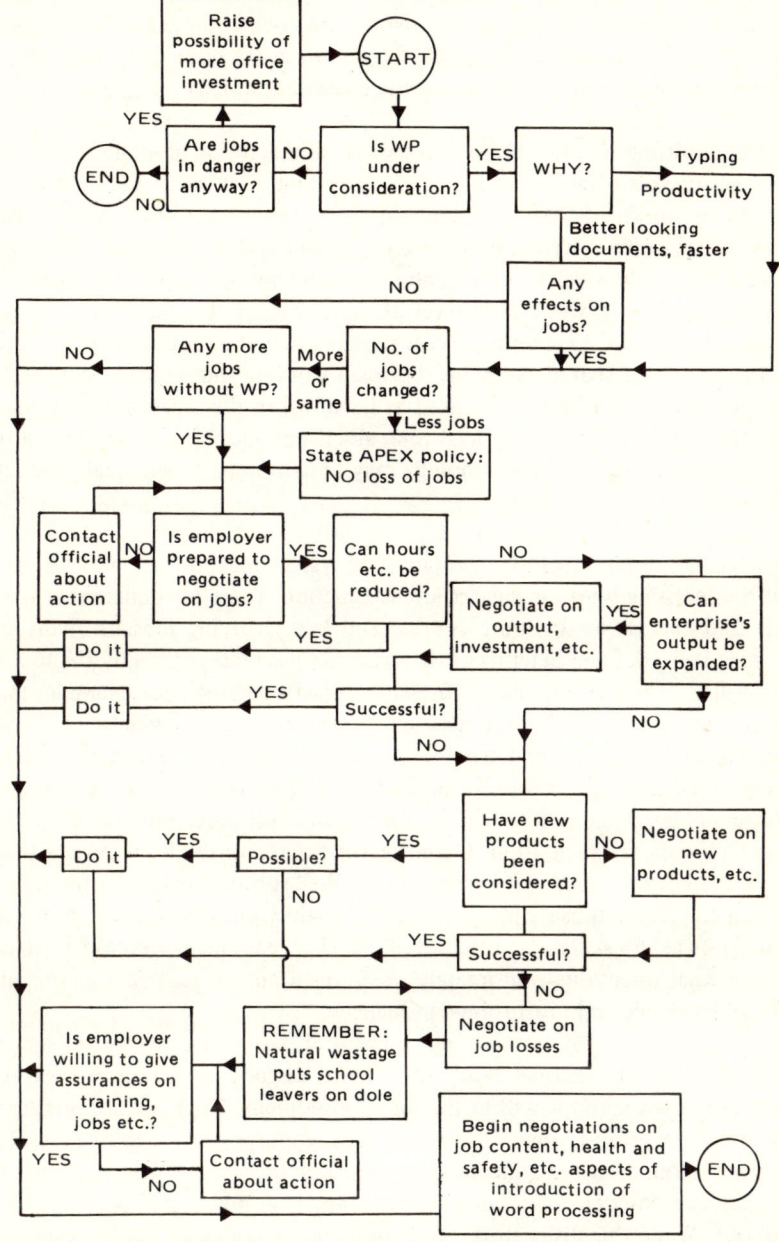

Source: APEX (1979), *Office Technology — The Trade Union Response*, London.

Figure 15.1. APEX strategy for the introduction of word processing and other office automation

preferences of unions for devolved industrial democracy could tend to push it in the latter direction. Information networks are compatible with those decentralised forms of organisation in which the groups at the nodes have relative autonomy—the mode of organisation has the possibility of becoming more horizontal than hierarchical.[32]

In permitting more complex control of manufacturing processes, information technology also offers unions a range of options relevant to the debate on the quality of working life. Experiments in less hierarchical forms of work organisations can involve, given union encouragement and approval, the redesign of production technologies to accommodate the need for job satisfaction and human dignity at work. Whatever the promise in the shape of spectacular productivity gains, long-term success requires that the productivity and quality control gains be shared equitably between workers and management, and that decision-making power be delegated as far down in the organisation as possible. Here the union occupies a crucial role, albeit one in which it may face a trade-off dilemma in which new technology may tend to improve the quality of labour while reducing the quantity.

Yet unions will necessarily retain conventional defensive functions. The transfer of routine tasks to a computerised system does not automatically involve more interesting work or greater job satisfaction. There is a danger that workers will find themselves in more routine and less satisfying jobs. In many cases, jobs are being defined in terms of the machines used rather than by the functions performed. The current use of the title 'VDU operator', for example, fails to define the large variety of clerical and administrative jobs which entail using a visual display unit. The tendency for computerised systems to take over relatively complex tasks which previously provided the opportunity for some job satisfaction could accelerate without joint control over pressures promoting specialisation and work fragmentation. Unions carry the responsibility to ensure that the majority of workers do not become a de-skilled substratum of machine-minders, and to emphasise those aspects of new jobs which provide variety, stimulation, new knowledge, skills and responsibility. The decision to develop technology in particular directions is ultimately determined not by technical factors alone, but by economic and institutional influences.

The assumption that technology is neutral, so commonly shared by union leadership at the national level, should not disguise the misgivings expressed by militant workgroups within the union movement. Their perceptions are such that technological change is seen as usually purposively deployed to lower relative labour costs and increase productivity (needlessly stripping employees of skill and choice), in the pursuit of competitive power and enhanced profitability.[33] While this dislocatory perception may occasionally induce resistance as a strategy among some workgroups (especially those faced with job uncertainty after displacement), the social costs of foregone productivity gains are often likely to be indirectly inflicted elsewhere in jeopardising the employment prospects for other segments of the workforce. The utilisation of existing bargaining power at the workplace to freeze work processes may be rational for some workgroups in the short run, but in the long run such resistance to change may equally prove self-defeating (as in the classic case of newspaper printers). There is no inexorable force determining either the logic of existing technology and workplace power relations, or the forms which the newer technologies should take during replacement processes. Militant shop-floor power is, therefore, faced with a paradox frequently obscured by political gesturing: mobilised workgroups can provide the foundation for securing the right to a negotiated share of technological change benefits, but they can also impede productivity gains through technological change in the first instance. There is more than one option for an alternative shop-floor strategy.

Although industrial conflict seems to be an integral characteristic of the industrial system, it is also a question of political choice whether the cost of restructuring the economy is borne mostly by the disadvantaged—by older workers who cannot retrain, migrants, minority racial groups, married women forced out of the labour market, and young unskilled workers unable to find permanent employment or adequate training facilities. Trade-union attitudes are already clear in demands for a joint voice in controlling technological progress. Concrete tactics are spelled out to most union organisations confronted with the introduction of new technologies—notably advance notice, consultation, full information upon employment effects, work environment, job organisation and training. But the direction of trade-union strategy will depend largely upon the societal successes or failures of broader economic, manpower and industrial relations policies in tackling the wider problems of expanding employment, in encouraging job mobility, and in providing sufficient new, better paid, higher skilled and more satisfying jobs to alleviate workplace insecurity.

Fear of unemployment (in an era when aggregate OECD unemployment approaches 30 million) is a powerful conditioned response to the prospect of technological change. Trade-union strategy will, therefore, look to governments to create a favourable expectational climate for structural employment changes. This would include a high degree of income support to workers displaced by new technology, encouraging transferability of pensions and accrued rights from job to job, legislating for initiatives to reduce working time, and adequate provision for early retirement. Unions will also urge that governments should accord high priority to employment creation, including public expenditure upon training in new skills, investment assistance for job creation, placement services and intensive research into the economic effects of different types of technology. But the social climate influencing industrial rights is equally important to the future pattern of union strategy. Unions, themselves an expression of the concept of industrial citizenship, believe strongly that radical structural changes in industry should not be legitimately allowed to take place outside the influence of democratic processes.[34] Government should encourage the extension of co-determination, consultation, negotiation and information rights of unions whose members are affected directly or indirectly by technological developments. This is not to imply that co-determination and consultation are justified only because of new technology, but the potentially vast effects of new technology underline the need to transform industrial relations so that employees may participate fully in decisions affecting their own jobs.

Overview

Forecasting is a perilous exercise. Yet, it seems certain that the new technologies will profoundly change industrial societies, although there are labour obstacles to be overcome (such as fear of redundancy and obsolesence of qualifications) and managerial obstacles (such as ignorance of technical advances, lack of risk capital and unwillingness to take risks in a high interest rate, recessionary environment).[35] However, there seems to be a fundamental consensus among unions that in the interest of economic growth, the wide application of new technologies should not be impeded. The structural changes in production are perceived by unions as likely to reduce the demand for labour, and hence unions insist that adequate measures be undertaken to cushion and overcome structural and individual effects. The information revolution in particular will also affect the supply side of the labour equation, tending perhaps to increase the quality of the labour force. Acting to reduce the supply of labour will be the increase in the diversity and opportunity of leisure activity made possible by the information revolution, leading to reduced work hours, more holiday time, early retirement and a withdrawal of people from the labour market. At the same time, the new technologies reinforce existing trends towards lifelong learning and training, and trade unions have been quick to recognise that initial education and training can only provide knowledge for starting positions in the future information society.

The historical significance of trade unions is central to modern society:

> The trade unions, defending and fighting for the interests of the working class, are responsible for its eventual incorporation into modern society, especially for an extraordinary increase in economic security, social prestige and political power.[36]

Trade-union opinion thus enfranchises more than the vagaries of fear, uneasiness or vested interest. It signals a social warning that it would be illusory to pretend that technological advances are without risks or negative consequences. Both positive and negative aspects are, and have been, part of man's work destiny. The unions concede that a critical attitude must be based on analysis and experience, and not expressed merely for the sake of opposing technology. But resignation to an all-powerful technostructure would be fatal. Whatever their shortcomings—and like all other labour market institutions, unions share knowledge imperfection—trade unions have consciously converged upon a classic problem in social philosophy.

The question is not so much whether we are the masters or the slaves of our machines, but whether machines still serve the world and its things, or if, on the contrary, they and the automatic motion of their processes have begun to rule and even destroy the world and things.[37]

Unions by themselves cannot provide more than a hazy and fragmentary answer. The manifold consequences of de-industrialisation and new technology[38] have yet to be recognised and mastered by all parties in the industrial relations system to provide an authentic answer for the next generation of post-industrial prophets.

Notes and references

1. Hence it would be invalid to claim that the nature of technology determines the organisational nature and attitudinal response of workers, just as it would be to claim that the sociopsychological characteristics of workers determine the manner in which a set of tasks is performed within a given technological matrix. The customs, expectations and preferences of workgroups are not merely orientations implanted into or inherited by the workplace. They are also influenced by the nature of the job and organisational structure.
2. See, for example, Wedderburn, D. and R. Crompton (1972), *Workers' Attitudes and Technology*, Cambridge, Cambridge University Press.
3. Even the Webbs in *Industrial Democracy* managed to disinter only two or three instances of open rejection of mechanisation. The President of the 1907 British TUC informed delegates that in relation to mechanisation: 'the proper course to adopt is to recognise the inevitable'. See Levine, A. (1967), *Industrial Retardation in Britain, 1880-1914*, New York, Basic Books. G. D. H. Cole insists that output restrictions in the form of ca'canny were discouraged by trade union leaders. See Cole, G. (1918), *The Payment of Wages*, London, Allen and Unwin, pp. 23-4.
4. OECD (1980), *Technical Change and Economic Policy—Science and Technology in the New Economic and Social Context*, Paris, p. 89.
5. The Employment Committee of the EEC, in 1980, revealed an interesting international portent during the discussion of the report on employment implications of the new microelectronics. While the workers' representatives (European Trade Union Congress) emphasised the need for full trade-union involvement in the introduction of technology, the employers (Union of Industry of the European Community) saw effective adaptation to technological progress as being conditional on respecting the 'independent' role of industry.
6. Although there is no consensus on the net job balance effect, most reports agree that robotics and numerical control machines will cause displacement in manufacturing, and that word processors will invade offices and displace clerks and secretaries. There is also some agreement on job destruction in specific segments of the labour market, with more severe displacement for older workers and less skilled workers. Women in information-processing functions in the service sector appear most at risk. Institute for Research in Public Policy (1979), *The Impacts of Computer/Communications on Employment in Canada—an Overview of Current OECD Debates*, Montreal. See also ILO (1982), *New Technologies: Their Impact on Employment and the Working Environment*, Geneva.
7. APEX has been one of the first British unions to analyse the effects of microelectronics with the view that, if current practices persist, the displacement effects will outweigh any job creation. A reduction is seen in the stock of routine manual and clerical

jobs, as a result of a combination of microelectronics, new computing systems and electronic telecommunications. The main impact is forecast for mailroom clerks, filing clerks, clerical and administrative jobs (involving large amounts of information, collation and transcription), cashiers, printing and reprographic staff, typists and secretaries, supervisors and front-line managers.

8. See Cockcroft, D. (198), 'New office technology and employment', *International Labour Review*, 119 (6): 689–704.
9. Trades Union Congress (1979), *Employment and Technology*, London.
10. Association of Professional, Executive, Clerical and Computer Staff (1979), *Office Technology—The Trade-Union Response*, London; APEX (1980), *Automation and the Office Worker*, London.
11. Association of Scientific, Technical and Managerial Staffs (1979), *Technological Change and Collective Bargaining*, London.
12. Banking, Insurance and Finance Union (1980), *Report of the Microelectronics Committee*, London.
13. Post Office Engineering Union (1979), *The Modernisation of Telecommunications*, London.
14. Confédération Française Democratique du Travail (1977), *Les Dégâts du Progrès—les Travailleurs Face au Changement Technique*, Paris.
15. The Swedish Confederation of Trade Unions (LO) issued a pioneering research report to the 1966 Congress. See Anderman, S. (ed.) (1967), *Trade Unions and Technological Change*, London, Allen and Unwin. A comparable report on data processing was made in September 1981.
16. European Trade Union Institute (1979), *The Impact of Microelectronics on Employment in Western Europe in the 1980s*, Brussels.
17. The Committee of Inquiry into Technological Change in Australia (1980), *Technological Change in Australia (Myers Report)*, Canberra, AGPS.
18. International Federation of Commercial, Clerical, Professional and Technical Employees (FIET) (1979), *Computers and Work*, Geneva.
19. International Metalworkers' Federation (1979), *Effects of Modern Technology on Workers*, Geneva.
20. Postal, Telegraph and Telephone International (1979), *Statement on Trade Unions and New Technology in Postal and Communication Services*, Geneva.
21. For Marx, technology, or the mode of production, constituted the *Unterbrau* which narrowly prescribed the whole superstructure of society, including class relationships and the existence and functioning of unions. For a wide range of social theorists, including Schumpeter and Durkheim, changes in technology constituted a critical factor producing tension among groups and classes, promoting industrial conflicts and providing the thrust for economic and social change. Yet, 'A kind of neo-Marxist model, in which technology is substituted for capitalism, has grown up in recent years, a model that has attracted a significant number of social scientists who seek to bypass the original Marxist concern about the economic formation of society. It is as if Marx both saw and overlooked technology . . . as if one can quietly retain his critique of capitalism while explicating assessments of the effects of technology and substituting the latter for the former. The effect—a kind of legerdemain—is to honour the dogma, in Joseph Schumpeter's fine phrase, by simply fingering the Marxist rosary.' Berg, I., Marcia Freeman and Michael Freeman (1978), *Managers and Work Reform*, New York, Free Press, p. 44.
22. See Cockcroft, D. (1980), 'New office technology and employment', *International Labour Review*, 119 (6): 689–704.
23. Mansfield, W. (1981), 'Technological change and the trade unions' in Evans, G., J. Reeves and J. Malbon (eds), *Labor Essays 1981*, Richmond, Victoria, Drummond Publishing, pp. 146–77.
24. See Jenkins, C. and B. Sherman (1979), *The Collapse of Work*, London, Eyre Methuen.
25. Yet it may be the case that firms have actually been slow to adopt more labour-saving technology since the cost of capital has risen, and it is the cost of labour relative to that of capital which is relevant to business decisions to install labour-saving technology.

Despite a rise in real wages, high interest rates (using Australia as an example) also raised capital costs to the extent that the cost of labour relative to capital remained near the long-run trend towards the end of the 1970s. Alternatively, the rise in real labour costs relative to capital in the mid-1970s may have been perceived as a short-run phenomenon, thus having little initial impact on production processes and substitution of capital for labour.

26. Lamberton, D. (1981), 'Social costs of technological change' in OECD, *Information Activities, Electronics and Telecommunications Technologies*, Paris, Vol. II, p. 206.
27. The substantive aims are given in the TUC's 'Ten Point Checklist' for negotiations: (1) No new technology which has major effects on the workforce should be introduced unilaterally; *status quo* provisions are recommended so that prior consultation is a condition for change. (2) The development of inter-union organisation and expertise. (3) Access to information before decisions are made. (4) Employment and output plans to be available. No redundancy if possible. Use of planned redeployment. (5) Provision for retraining with improved or maintained earnings. (6) Reduction of working hours. (7) Pay structures to be preserved, subject to a move towards 'single status' and 'equal conditions'. (8) Union to have influence in system design and new equipment not to be utilised in work measurement. (9) Observance of health and safety guidelines. (10) Establishment of review procedures.
28. A wide appraisal, from which much information has been gathered, is Manwaring, A. (1981), 'The trade-union response to new technology', *Industrial Relations Journal*, 12 (4): 7-26. The major areas of 'model agreements' are: procedure, security, job content, benefits, health and safety.
29. Change in recruitment catchments is occurring, but no conclusive judgements can be made upon the unionisation effects of new technology; most models yield exceedingly complicated and conditional predictions. It is probable that white-collar union penetration rates will increase. Aggregate composition changes by sector, age, sex, skill, etc. appear indeterminate in the list of superimposed recession cycles.
30. Mansfield, W. (1981), 'Technological change and the trade unions' in Evans, G., J. Reeves and J. Malbon (eds), *Labor Essays 1981*, Richmond, Victoria, Drummond Publishing, pp. 146-77. Australian unions number more than 300.
31. Cited in Forester, T. (ed.) (1980), *The Microelectronics Revolution*, London, Blackwell, p. 386.
32. Brooks, H. (1981), 'A critique of the concept of appropriate technology', *Bulletin of the American Academy of Arts and Sciences*, 34: 16-37.
33. See Transnational Co-operative (1978), *The Job Killers: Technology and the Economic Crisis of Workers*, Sydney.
34. For example, without exception, 30 bank unions affiliated to FIET in Europe called for the acceptance of change by negotiation and consultation, and for more information on capital planning. Some unions stated a desire to oppose the introduction of further computerisation if these assurances were not met. The BIFU 1980 Annual Conference declared that 'unless such co-operation is forthcoming this union will actively oppose any management attempts to unilaterally introduce such technology within the banking and finance industry'. See also FIET (1980), *Bank Workers and New Technology*, Geneva.
35. For a wide bibliography (over 450 references) on the relationship between technology and the workforce in western industrialised societies, see Ford, W., M. Coffey and D. Dunphy (1981), *Technology and the Workforce: an Annotated Bibliography*, Sydney, Technology Research Unit, Government of New South Wales.
36. Arendt, H. (1959), *The Human Condition*, New York, Doubleday, p. 193.
37. Ibid., p. 132.
38. To simplify the issues, the term 'new technology' has been used in this chapter to designate chiefly microelectronics technology (especially miniaturisation and large-scale integrated circuits) and the information-intensification process. The generational development of semiconductor technology has produced a hybrid: 'mechatronics' (a combination of mechanics and electronics), with applications in robotics. Developments in informatics, new materials (carbon fibres, fine ceramics, high polymer

conductors, etc.), biotechnology, energy and other fields, are all parts of the new technology revolution. Traditionally, trade unions have regarded technology as the application of scientific knowledge in technical systems and technical methods of production of different products. The union term 'techniques' was used to designate the way in which raw materials, machines, equipment and plant, production process, units of production and products are utilised and combined.

16. The trouble with techno speak

Ian Reinecke

The trouble with ignorance about technology is that it is so seldom admitted by the ignorant. The intimidatory effect which the use of technological terms has had on the non-technical community has been devastating. It has reduced many otherwise intelligent, articulate, knowledgeable people to silence on matters technological. Where the technical language concerns computers and electronic technology generally, the silence extends beyond not knowing to that more dangerous phase of not being game to ask. There have been too few protests by the non-technical at the tyranny of technological language about computers. Why is it that we have not heard more of this sort of heart-felt outburst from an American trial judge, after listening to evidence in a computer case?

> ... the first finding the court is constrained to make is that, in the computer age, lawyers and courts need no longer feel ashamed or even sensitive about the charge, often made, that they confuse the issue by resorting to legal 'jargon', law Latin or Norman French. By comparison, the misnomers and industrial shorthand of the computer world make the most esoteric legal writing seem as clear and lucid as the Ten Commandments or the Gettysburg Address; and to add to this Babel, the experts in the computer field, while using exactly the same words, uniformly disagree as to precisely what they mean.[1]

The first point to be made about the language employed about computers is that its use is not accidental. At least some of the reasons for its existence are shared by other groups in society. As the judge points out, lawyers have a professional language barricaded to outsiders by Latin terms which are perfectly comprehensible to those who practise law. To the lay public, however, many of the terms used in the courts are meaningless. Medicine employs a similar professional language, intended for internal consumption, which effectively excludes outsiders. The law and medicine are probably the most usual points of contact between ordinary citizens and professional languages. There are many languages operating in the less familiar environment of physics, chemistry and biology, and most of the other branches of science. In applied science, there are specialised languages in civil, electrical, marine, aeronautical and other forms of engineering. The characteristic shared by law, medicine and computing is that all three operate in a way which brings the general public into touch with them. People need lawyers from time to time and they require medical advice. Increasingly, they are being brought into contact with the world of computing.

The users of professional languages defend them on the ground that they facilitate communication among those who work in the profession. Many of the terms are used as names. Describing a legal principle or a medical condition would be far more laborious if, each time it was referred to, a full explanation was required. Substituting technical terms allows professional discourse to flow more freely. Around those describing terms and phrases, linguistic conventions flourish—as anyone who has watched polite court room disagreements between opposing barristers will testify. Taken together, the technical terms and the polite conventions of the professions effectively exclude lay audiences from much of the discussion of lawyers, doctors and other professionals.

This exclusion is less critical in areas where the public is not involved in daily contact with technical areas. Ordinary citizens tend to concern themselves with matters like pure physics only when there is an issue of social importance at stake; often a lack of detailed understanding of physics is not an impediment to rational discussion about the results of the application of its principles. The widespread debate about nuclear weapons does not require a physicist's understanding of how the physical reactions work in atomic or hydrogen bombs. Working explanations of what is involved when nuclear bombs explode is sufficient for the question of their use to be discussed. The explanations have been given wide publicity since Hiroshima and Nagasaki, and although the definitions used by the media may be inadequate for physicists, they are adequate for the purposes of public discussion. The fact that nuclear weapons have become a matter for public concern has forced some translation of the technical language of physics.

Sometimes that pressure is not required to make technical terms comprehensible. In botany, for example, the technical language of description has always been mirrored by a common language. The technical language of botany was preceded by the descriptions of trees, flowers, shrubs, vines and vegetables which evolved through people's everyday existence. Before Joseph Banks, the botanist accompanying Captain Cook on his voyages to Australia, set foot on land, the country's flora were already described in Aboriginal language. Every suburban gardener knows that there is a botanical name for the tree known commonly as the silky oak or the shrub called Christmas bush. To a certain extent, the same thing occurs in biology, where lay terms have their equivalents in the technical language used by biologists. In the case both of nuclear physics, where working descriptions have arisen as a result of public discussion, and botany and biology, where there was already a language in common use, the exclusion of ordinary people is less critical than it threatens to be in electronics.

Where there is no bridge between the understanding of lay people and technocrats of whatever persuasion, exclusion is inevitable. Where common explanations have not emerged or where ordinary language is not a mirror of a technical vocabulary, there are other means by which the gap in comprehension can be crossed. In the courts this occurs in a formal manner. The existence of the jury system compels lawyers and judges to explain legal points in ordinary language. The compulsion on the police to explain people's rights, no matter how imperfectly practised, at least recognises that the public left in ignorance could forfeit those legal rights. In medicine, the need for doctors to obtain a patient's official consent for certain forms of treatment requires an explanation which does away with the medical descriptions.

The single most worrying feature of the use of a language different from that of ordinary people is that no means of translation is readily accessible. Some translation is provided in a variety of ways in the law, medicine, physics, botany, and biology. In the area of computers and electronics technology, there is no such bridge for the understanding of those who are non-technical. Seldom is the effort made by practitioners of electronics to explain in plain language what they are doing, as nuclear physicists have been obliged to do. There is neither the formal requirement which compels lawyers to explain themselves in ordinary language, nor any practical necessity, as is sometimes the case with doctors. In the absence of those requirements, electronics remains a subject largely inaccessible to the general public. At the same time, the intrusion into their lives of this technology increases daily. The conjunction of inaccessibility of the language of computing and the pervasiveness of the technology creates a difficult problem indeed for anyone wanting to discuss the issues surrounding electronics.

More than anything else, the barriers to understanding high technology have contributed to an absence of public discussion about what its effects will be. Although good minds have expressed themselves on issues such as nuclear technology, the legal system, modern medicine and other areas where technical language is used, they are largely silent about electronics. Even the best minds need a bridge from their world to that of technology, science and the professions. In electronics, the strands which might together form a bridge are widely dispersed and by themselves insubstantial. The problem for many people educated in a non-technical or humanist tradition is how to enter the discussion. Sources of information to enable that entry are not easily accessible, but they are there.

Starting at the least accessible end of the lode of information which exists about electronic technology, there are the formal courses and textbooks used in any technical area. It should be no more necessary to undertake a diploma in electronics in order to understand the social effects of its deployment in machines like computers than it is to qualify in the law to discuss legal issues. Because the bridges to understanding electronics are so fragile, some comprehension of technical terms is required. The specialist media offer one strand to bridge the gap, although often in an unsatisfactory manner. Consumer electronics magazines, which address themselves to enthusiasts, seldom deign to explain themselves to those who have failed to crack the technological codes involved. Similarly, professional electronics magazines employ the language of electronics engineers, computer programmers and equipment designers, with little explanation to lay readers. In most newspapers, articles on electronics technology which do more than endorse the latest claim of computer manufacturers are rare. In this regard there is little to distinguish the serious newspapers from the popular tabloids. The treatment by the media of the effect of electronics technology on work has been scant, and treatment of its impact on people's private lives has been even less substantial. The same can generally be said for television, a medium which is itself the creature of high technology. With the exception of radio documentaries, there has been little attempt to bridge the gap between lay and technical audiences. As a result, the use of those sources has been a rather catch-as-one-can matter.

In general, the books written for lay audiences about technology have been the main source of information for non-technical readers. There have been a number of books published which attempt to explain electronics in ordinary language, with different degrees of success. Few have attempted to mix explanation with any critical appraisal of the technology, and some are unabashed hymns of praise to electronics and computers. In Britain, in particular, many of these books have been written by computer professionals or by journalists who cover electronics in their full-time occupations. Their books tend to reflect the uncritical stance of both the professionals working with computers and the press which reports them. Only the emergence of heretics promises to break this comfortable hegemony.

Between them, these sources can yield some of the information necessary for non-technical people to understand computers, and hence the issues that they produce. In addition, there are at least three groups of people who do understand the technology and who represent different interests in the production, sale and consumption of computers and electronics systems. They are the designers and builders of the machines themselves, the people who sell them, and the people who use the equipment. Of these, the people in the first group are the most obvious users of language which excludes public discussion: the builders of computers, like physicists, need shorthand descriptions and abbreviated terms to describe components and processes. The technical language of computing and electronics is littered with more acronyms than the language of any other discipline, and one suspects that the existence of the first acronyms created the environment for more. The names of the computer suppliers—IBM, DEC, CDC, DG, HP and ICL—are really International Business Machines, Digital Equipment Corporation, Control Data Corporation, Data General, Hewlett-Packard and International Computers Limited. They are seldom referred to by their full names in the computer industry. The various machines produced by those companies are also referred to in code, although generally in numbers rather than letters. The components of the machines are also abbreviations. A RAM is a random access memory, a ROM a read only memory, an EPROM an electrically programmable read only memory, and an EEPROM an electrically erasable programmable read only memory. The storage capacity of magnetic tapes or discs is quantified in thousands of bytes of memory, as in discs with a capacity of 64K bytes. The types of micro chips available are described as MOS circuits, for metal oxide on silicon, or CMOS for complementary metal oxide on silicon.

It is not this language alone with which most people come into contact when computers are discussed. The basic technical terms and definitions which may be employed with some precision by electronics engineers are the linguistic raw materials for those who sell computers and their components. To those terms are added words which are sometimes taken from ordinary language, but which are used in a new sense by the marketers of electronic equipment. 'Interface' is one, meaning the connection between two things. 'Modular' is a ubiquitous term in computer marketing, and it refers only to the fact that units can be added to a computer system. 'Flexibility' is a word much favoured by the sales forces of computer companies; it means only that the machine can be used for a number of purposes. Other indicators of the possibility of machines being used for more than one task are words such as 'multi-user' and 'multi-purpose'. Where a machine operates on its own, it is said to be 'discrete', or that it works as a 'stand alone' unit. If one machine can function connected to another, they are said to be 'compatible'. If they are supposed to be simple to operate, they are called 'user friendly'. Even at the most general level, programs are likely to be called 'software' and machines 'hardware'. There is even the term 'firmware' to describe a machine which incorporates some programming.

If none of this sounds too difficult for a lay audience to grasp, consider the conjunction of technical and marketing terms which is used in the computer industry. Let us say we are examining the merits of a word processing (WP) machine and are confronted with a salesperson who says:

> This is a stand alone WP with multi-purpose capability either as a discrete unit or by an interface giving additional flexibility and compatibility to other units by a modular upgrade path and user friendly software. This state of the art of WP has 64K of RAM, can be used in BASIC or COBOL and is IBM's top of the range Z1000 series.

As an example of the genre, this ranks among the more comprehensible. When the marketers of computers are not striving for such clarity, their language becomes even more dense. Without the ability to translate the above, the person on the receiving end would not be sure what it was he or she was being asked to buy. For the non-technical person interested in understanding computer technology, the gap in understanding may be even wider. It is the marketing departments of computer companies which have formed the language to describe the technology. They have taken the technical terms out of the design laboratories and set them loose in a jungle of phrases and words which define the world of computing and separate it from that of ordinary discourse. When the technocrats of electronics complain that their terms are used imprecisely it is worth remembering that their language has been usurped by the people who sell the goods they produce.

The third group of people familiar with the language which has grown up around computers and electronics consists of those who buy and operate the technology, most often referred to as the end users. They generally receive their initiation into the new language from the marketing staff of suppliers, rather than the designers. That education process may not be entirely voluntary. In any selling process, the salesperson's task is eased considerably if he or she is operating with a home ground advantage. One of the most valuable assets is being able to control the language in which the discourse occurs; the technique is referred to as 'baffling 'em with science'. The technical language of the computer industry is one of the most sophisticated sales aids ever devised.

Choosing a computer could be compared with buying a new motor car. Each prospective customer may have identified a group of reasons that are going to be used in deciding which car to buy. They may include price, performance, colour, accessories, degree of maintenance, reliability and ease of driving. Discussion with the car retailer may include phrases such as fuel injection, independent suspension, turbo-assisted, and radial tyres. Because buying a car is an activity most people engage in at some time or another, these otherwise technical phrases have entered everyday language. That is not the case when the purchaser is considering which computer to buy.

The success of minicomputers, which now represent a considerable proportion of the total computer market, is very much a story of how marketing broke through ignorance and fear in the business community. It was assisted in doing so by the development of a language which gave computer buyers a way of describing the machines they bought and their functions. Often, it gave them little else. Many purchasers of minicomputer systems, especially in small companies, swallowed the marketing line fashioned so assiduously by computer suppliers. There were many unsatisfied customers as a result. Like buying a car, purchasing a computer requires some independent assessment of the seller's claims. In the case of the people who sold computers, they had the double advantage of appearing both as salesperson and as expert. In some cases, that expertise consisted of not much more than the ability to master the technical terms, the key phrases, the buzz words and the jargon of the computer industry. Afraid to question such obvious competence, and unable to check the claims made by the computer sales staff, company owners and managers bought and repented at leisure.

One of the main reasons why purchasing mistakes were made was the lack of any description and evaluation of the technology which was independent of the hybrid technical and marketing language used about computers. When it became clear that this was the reason for many poor computer-purchasing decisions, there was a demand in the business community for information and education. The computer suppliers recognised that the advantages in the gap in understanding were only short term, and began the process of educating buyers through user groups and educational seminars. In addition, consultants were engaged to assist in computer-buying decisions, and specialist newsletters sprang up to assist purchasing decisions. The net result has been that users of computers speak the language they have been brought up in—the language created in the marketing department of their suppliers.

This has lessened demand for computer terms to be translated into layman's language and, combined with managerial elitism, has strengthened the ties between supplier and user. An important reason for computer purchasers in industry and commerce taking up their suppliers' language is that it serves their own purposes as well. It is commonly perceived by users of computer systems in business to be in their interest to keep the workforce uninformed about technological developments. They have been assisted in this by the adoption of a language which exludes their non-technical staff from meaningful discussion about the technology. Despite the dire predictions of the priests of high technology, there has been relatively little resistance from the trade unions to the introduction of computers at work. One of the reasons, according to the unions, is cursory consultation about the introduction of the technology. It is introduced before the workforce or union representatives have been able to evaluate its effect. The unexpected source of resistance has come from middle management, which does have access to more information than the shop-floor, office or factory worker.

The source of information for middle management is quite often the data-processing departments of large companies, normally the repository for what could be considered the shock troops of computerisation. The creation of these departments is necessary for computers to be introduced, but at the same time they may endanger upper management by releasing information into the workplace. The technical expertise such staff as programmers possess is allied with managerial status within many organisations. They are able to foresee the technological options open to management, and to rank their preferences according to different criteria, such as job enhancement, improved career opportunities, and enlarged data-processing departments. Programmers and systems analysts often make it easy for their employers to isolate their expertise from the rest of the workforce. Physically, they often work in computer rooms separated from the rest of the staff, but an equally important factor is the language they speak, which makes it difficult for other employees to discuss the technology on common ground. So while computer professionals may provide an occasional alternative source of information for other staff, their own self-interest may keep them from helping non-computer professionals devise ways of minimising the effect of electronic technology on the work they do. The very nature of their work means that the computer staff will be generally better educated, younger, more highly paid, and more homogeous than other groups of workers.

The role of filtering information to the workforce on matters technological falls to selected staff known in the handbooks of the computer suppliers as 'facilitators'. Their job is to facilitate the introduction of computer technology by convincing the rest of the workforce that it is in their interests not to oppose it. Facilitators have unkindly been likened to the sheep used in slaughteryards in rural Australia called bell wethers. They are old sheep, wearing bells around their necks, which lead flocks through chutes to the slaughterhouse. While the bell wethers pass through unharmed to repeat the process with the next flock, the sheep which follow are slaughtered. Computer suppliers stress the importance of recruiting the right sort of facilitator, with the emphasis on such qualities as leadership. The facilitator is taught the language of the designers, sellers and users of technology in order to exhibit expertise to the rest of the workforce.

These elaborate means are aimed at spiking trade-union opposition, or even unorganised hostility, to changes in work processes, loss of jobs and erosion of skills which occur when computer technology is introduced. By denying workers information about the technology, management achieves a significant advantage —it is able to determine the point at which discussion about computer systems will begin.

By refusing to divulge information, the employer can select the basis of discussion with the unions so that it focuses on how best to implement a system of the employer's choosing. Computer suppliers represent their products as solutions to problems. Definition of the problem, and decisions about its solution, occur between computer vendor and the user. After the selection is made, the unions are notified rather than consulted. This typical pattern of the way in which computers are introduced means that the unions are reduced to arguing about the effects of a particular machine or system. Were they able to be involved in discussions about the problem the computer has been introduced putatively to solve, management would have significantly less freedom to come to agreement with suppliers about the purchase of particular machines. It is also in the supplier's interest to exclude workers from discussions; while employers are unlikely to question the fundamental assumptions which are designed into their machines, employees may well have cause to do so.

When computer systems for offices or factories are designed, not only technical parameters are observed. The system must satisfy certain requirements which are laid down by the marketing departments, or arrived at by studies of what sort of machine will appeal to the market. These criteria are based on what is perceived to be management's desire for more control over the workplace, greater detail in reporting within hierarchical structures, more information about work processes, increased monitoring of staff, and increased capacity to spot non-standard performances. Other requirements might be more standardisation in work practices and more analysis of the constituent parts of a particular process, whether it be paying an invoice or turning a metal component on a lathe. The machine which is designed by the computer supplier must build in these assumptions because they are some of the primary reasons for buying the equipment in the judgement of the marketing department.

While the needs of management, whether real or imagined, are catered for in the design of computer systems, the sometimes conflicting desires of the workforce do not even reach the drawing board. It is theoretically possible to design office computer systems which increase individual autonomy over work. There is no reason why computer technology should not be introduced to make a more interesting job for the human being who has to work with it. But these considerations are unlikely to be advanced by computer company sales staff as a reason for the system's purchase by employers. Releasing information at the bottom level of the pyramid threatens the whole structure. For that reason, there is a positive disincentive for suppliers to teach anyone but the management hierarchy, to whom they sell, the language of computing and electronics.

We began with the American trial judge's outburst against the tyranny of language exercised by the computer industry at all levels. There are clear economic interests challenged by the demand that the computer technocrats translate what they are doing into ordinary language. It is best to recognise this at the outset of any journey to discover what is happening in the world of computer technology. Instead of proving a barrier to investigation, as is partly intended, the opacity of the language of electronics should be a challenge to the non-technical inquisitor. The only way to information involves transcending that barrier, and the best strategy is, as it always has been, the relentless posing of difficult questions. That is a worthy task for those who believe that exclusion from public debate of technological matters leads inevitably to dangers which threaten such basic concepts as freedom and democracy.

Notes and references

1. Quoted in Wessel, M. (1975), *Freedom's Edge—The Computer Threat to Society*, Reading, Mass., Addison-Wesley, p. 127.

17. The difficulties of national innovation policies
Roy Rothwell

Introduction

One difficulty in establishing national innovation policies lies in first establishing a generally accepted definition of what precisely constitutes an 'innovation'. When we talk about the process of technological innovation, do we mean the advance of technology on a broad front, or do we mean the process of innovation at the level of the firm? Does an innovation need to contain great technical novelty (a major innovation), or does an improvement to an existing product or process (incremental innovation) qualify? It would seem sensible for national innovation policies to include all these.

Not only is there no universally accepted definition of innovation, there is currently no satisfactory 'theory' of innovation either. It is, perhaps, Nelson and Winter who have come closest to constructing such a theory.[1] Fortunately, despite the absence of a general theory of innovation, there now exists a considerable—and rapidly growing—body of empirical knowledge concerning technological innovations on which national policies can be founded.

The next question to ask is 'what is innovation policy?' The answer, for the purposes of this chapter, is that it is essentially a fusion of science and technology policy and industrial policy, both of long standing. The former consists of the patent system, technical education and the promotion of basic and applied research within the scientific and technological infrastructure; the latter consists of such measures as taxation policy, investment grants, tariff policy and industrial restructuring. Today it seems rather obvious that the two should be closely integrated, which is reflected in the recent spate of official national innovation policy statements.

If we accept the above definition of innovation policy, we can then go on to describe and to classify a set of governmental innovation policy tools (see Table 17.1). These tools can be subdivided into three broad categories:

(i) supply side tools — these include the provision of financial and technical assistance, including the establishment of a scientific and technological infrastructure;

Table 17.1. Classification of government policy tools

Policy tool	Examples
1. Public enterprise	Innovation by publicly owned industries, setting up of new industries, pioneering use of new techniques by public corporations, participation in private enterprise
2. Scientific and technical	Research laboratories, support for research associations, learned societies, professional associations, research grants
3. Education	General education, universities, technical education, apprenticeship schemes, continuing and further education, retraining
4. Information	Information networks and centres, libraries, advisory and consultancy services, data bases, liaison services
5. Financial	Grants, loans, subsidies, financial sharing arrangements, provision of equipment, buildings or services, loan guarantees, export credits, etc.
6. Taxation	Company, personal, indirect and payroll taxation, tax allowances
7. Legal and regulatory	Patents, environmental and health regulations, inspectorates, monopoly regulations
8. Political	Planning, regional policies, honours or awards for innovation, encouragement of mergers or joint consortia, public consultation
9. Procurement	Central or local government purchases and contracts, public corporations, R & D contracts, prototype purchases
10. Public services	Purchases, maintenance, supervision and innovation in health service, public building, construction, transport, telecommunications
11. Commercial	Trade agreements, tariffs, currency regulations
12. Overseas agent	Defence sales organisations

Source: Rothwell, R. and W. Zegveld (1981) *Industrial Innovation and Public Policy*, London, Frances Pinter, p. 161.

(ii) demand side tools — these include central and local government purchases and contracts, notably for innovative products, processes and services;
(iii) environmental tools — these include taxation policy, patent policy and regulations (worker health and safety, environmental and economic).

Some recent national innovation policy formulations

Before going on to discuss the difficulties of national innovation policies, some fairly recent policy formulations in several countries will be briefly described. The analysis is based on the reports listed in Table 17.2.[2]

A commitment to the formulation and implementation of innovation policies does, of course, imply a belief, by governments, in the importance to their countries of enhanced rates of technological change. Certainly there is a growing belief amongst western governments that innovation is, partially at least, one means of breaking out of the current recessionary cycle. Specifically, innovation is seen as enhancing international competitiveness, as contributing to and, in some instances, even as driving economic growth, and as a potent means of generating new jobs. Figure 17.1 shows some of the benefits of industrial innovation as reflected in the policy statements listed in Table 17.2.

Table 17.3 lists the main tactical objectives of the innovation policies of six major governments (Canada, Japan, the Netherlands, Sweden, the UK and the US), and Table 17.4 lists the corresponding set of main policy measures suggested in order that these tactical objectives might be met. Table 17.5 presents an analysis of policy recommendations in these countries by type of policy tool.

Analysis of the tables highlights a number of important differences between countries. For example, in Canada, Japan and the Netherlands, the most favoured tool is the 'scientific and technical' one, for the UK 'financial and taxation', and for the US 'legal and regulatory'. In the case of the US, this reflects deep concern that the economy is over-regulated.[3] In both the UK and the US, policy appears to be focused mainly on getting the overall climate for innovation right, while in the other countries the focus is more on specific tools, mainly on the supply side.

Perhaps the most important difference lies between those nations (for example, Japan and, to a lesser extent, Sweden and Canada) that have clearcut, long-term strategies towards the development and exploitation of specific product groups and new technologies, and those that do not. The former nations appear to have accepted that certain structural changes have taken place in the world economy that must be mirrored by changes in their own economies; that greater advantage is to be gained from exploiting changes in the new world economic order than from steadfastly resisting those changes through measures

Table 17.2 List of reports used for analysis of national innovation formulations

Country	Report
Canada	*Forging the Links: A Technology Policy for Canada* Science Council of Canada, Report 29, Ontario, February 1979.
Japan	*The Role of Technology in the Change of Industrial Structure* (abstract) Industrial Research Institute, Japan, April 1978.
Netherlands	*Summary of the Government White Paper on Innovation* Science Policy Information Department, The Hague, 1979.
Sweden	*Technical Capability and Industrial Competence: A Comparative Study on Sweden's Future Competitiveness* IVA Royal Swedish Academy of Engineering Sciences, Stockholm, June 1979.
United Kingdom	(1) *Industrial Innovation* Advisory Council for Applied Research and Development (ACARD), London, 1978. (2) *Technological Change: Threats and Opportunities for the United Kingdom* ACARD, London, December 1979.
United States	(1) *The US Domestic Policy Review on Industrial Innovation* (interim report), May 1979. (2) *Advisory Committee on Industrial Innovation, Final Report* United States Department of Commerce, Washington, September 1979. (3) *The President's Industrial Innovation Initiatives* (fact sheet), Washington, October 1979.

Source: Rothwell, R. and W. Zegveld (1981), *Industrial Innovation and Public Policy*, London, Frances Pinter, p. 56.

seeking to protect ailing industries. In contrast, other countries (for example, the US) have left technology choice largely in the hands of private companies, the direction of technology being determined by more immediate market forces.

At a more general level, we can state that another major difference between countries lies in their very approach to innovation policies, which reflects differences in the role that governments play in the economy and in industrial development generally. In this regard, we can discern two kinds of state intervention. In some countries state intervention in industry is seen as a major part of a process of indicative planning. This is the case in such countries as France and Italy, where industrial policy is used as an important instrument for economic policy and where the objectives of that policy are formulated within a framework of economic and social development plans, which are indicative for the private sector. Industrial (innovation) policy is then formulated through consultative and co-ordinative procedures and institutions within government and between government and industry.

In other countries industrial innovation policy is seen as simply one component of general economic policy, aiming to create a favourable climate for

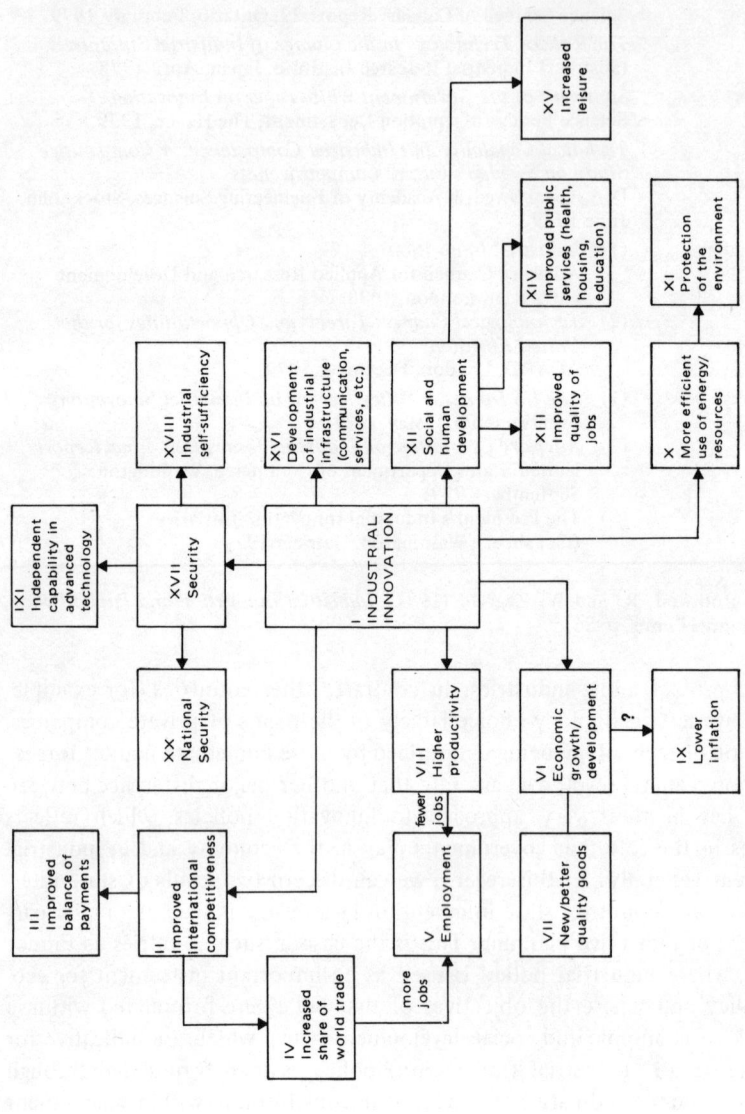

Source: Rothwell, R. and W. Zegveld (1981), *Industrial Innovation and Public Policy*, London, Frances Pinter.

Figure 17.1. Industrial innovation—possible benefits

Table 17.3. Industrial innovation policies—tactical objectives

Canada	Japan	Netherlands	Sweden	United Kingdom	United States
(a) Increased demand for Canadian technology	(a) Gradual change in industrial structure	(a) Increased capacity for innovation in industry	(a) Improved performance in traditional sectors	(a) Strategies for industrial sectors	(a) Improved technology transfer
(b) Expand Canadian industry's potential to develop technology	(b) Identification of future areas of growth in industry	(b) More government R & D	(b) Strengthen the knowledge base	(b) National policy to co-ordinate, for example, employment	(b) Increased technological knowledge
(c) Strengthen capacity of firms to absorb technology	(c) Construction of innovation policies after public consultation	(c) Better consultation and information services	(c) Identify and exploit new technological prospects	(c) Increased technological research	(c) Improved patent system
(d) Import technology under favourable terms		(d) Increased demand for innovation to satisfy public needs		(d) Better consultation and information services	(d) Improved anti-trust policy
				(e) More/better trained manpower (especially engineers)	(e) Foster development of small innovative firms
					(f) Improved federal procurement
					(g) Improved regulatory system
					(h) Facilitate labour/management adjustment to innovation
					(i) Supportive attitude to innovation

Source: Rothwell, R. and W. Zegveld (1981), *Industrial Innovation and Public Policy*, London, Frances Pinter, p. 69.

Table 17.4 Industrial innovation policies—main policy measures

Canada	Japan	Netherlands	Sweden	United Kingdom	United States
a.1 Innovative-conscious procurement	(Not specified in English abstract version)	a.1 Help large firms meet costs of R & D	a.1 Broaden markets	a.1 Mature industries	a.1 Information centre on federally supported R & D/technology
a.2 Major government programmes		a.2 Ensure small firms use advantages (for example flexibility)	a.2 Standardisation	(a) import technology	a.2 Foreign technology information
b.1 Sponsor companies		b.1 Reorientate Dutch R & D to social and industrial problems	a.3 'Niche' strategies	(b) co-ordinate public and private sectors	b.1 Generic technology centres
b.2 Encourage consortia and joint ventures		b.2 Research in new areas of technology	a.4 Search for new markets (for example, in developing countries)	a.2 'Laggard' industries	b.2 Regulatory technology research
b.3 Aid to small firms		b.3 Better use of existing expertise	a.5 Exploit national advantages (for example, resources, technical competence)	(a) R & D batch production	b.3 Improved university-industry co-operation
c.1 Sectorally orientated technical centres		c.1 Co-ordination of consultation and information services	b.1 Improve school curricula	(b) monitor technological changes abroad and import	c.1 Uniform government patent policy
d.1 Secure maximum advantage for Canada from imported technology		d.1 Innovation-conscious procurement	b.2 Improve university courses	a.3 Service industries	c.2 Improved patent service
d.2 Co-ordination of importing of technology so as to foster selective development of Canadian technological capability		d.2 Innovation-conscious regulation	b.3 More technical and scientific research	(a) more R & D	d.1 Funds for small firms
			c.1 Funds for research in new technologies	a.4 New industries	d.2 Corporations to sponsor innovation (like NRDC)
				(a) co-ordinated national strategy	d.3 Increased venture capital
				a.5 Small firms	e.1 Uniform procurement policies
				(a) better consulting services	f.1 Develop a technology forecasting system
				(b) better technology transfer (for example from large firms)	g.1 Award for innovation
				(c) aid for R & D	
				b.1 Better university-industry liaison	

Source: Rothwell, R. and W. Zegveld (1981), *Industrial Innovation and Public Policy*, London, Frances Pinter, pp. 70–1.

Table 17.5 Analysis of policy recommendations by type of tool

Type of tool	Canada	Japan	Netherlands	Sweden	United Kingdom	United States
1. Public enterprise	0	0	0	1	1	0
2. Scientific and technical	7	7	9	3	4	4
3. Education	3	1	5	11	4	3
4. Information	2	2	8	2	3	8
5. Financial	5	2	6	5	6	4
6. Taxation	1	0	0	1	6	13
7. Legal and regulatory	0	0	6	1	0	46
8. Political	2	4	2	3	4	2
9. Procurement	4	0	2	2	3	11
10. Public services	0	0	1	0	3	0
11. Commercial	2	1	1	0	0	3
12. Overseas agent	0	0	0	1	2	0
Total	26	17	40	30	36	94

Source: Rothwell, R. and W. Zegveld (1981), *Industrial Innovation and Public Policy*, London, Frances Pinter, p. 73.

industrial development. Although these countries, like the Netherlands, Denmark and the Federal German Republic, use industrial (innovation) policy instruments, or even sectoral policies, these policies are not formulated within the framework of a National Plan, nor are they used as selective policies in an intensive or systematic way. This distinction between two ways of formulating innovation policies should not, of course, be seen as a model for describing two totally different worlds, since differences are often not as great as they appear. Other differences between countries may be more important.

Some problems with innovation policy

Taking first the question of innovation subsidies, a number of economists have questioned the actual need for direct government subsidies for industrial projects, arguing that this will ultimately lead to an intervention-breeding system as industrialists seek profits increasingly through political lobbying and through subsidies, instead of through producing and selling market-orientated goods.[4] In other words, government subsidies that are too easily obtainable will result in projects of low market potential, and perhaps over-sophistication, introduced by major companies feather-bedded through public subsidy.

It might be, for example, that government subsidy, linked to military procurement, has been a main cause of British semiconductor companies having exhibited a high degree of technological virtuosity and a low level of linkage to the civilian marketplace. The combination of subsidies and procurement has resulted in a preponderance of 'technology-push' and a paucity of 'need-pull' in the operation of some major UK semiconductor companies.[5]

Certainly, there exists evidence from several countries to suggest that government funds have too often gone to support projects of high technical sophistication, but of low market potential and profitability.[6] As well as having lower market potential, projects subsidised by governments have often also involved higher technical and financial risk than those funded wholly by private companies. Walker, for the UK, has indicated an inverse relationship between scale, technical sophistication and government involvement on the one hand, and commercial success on the other.[7]

The fact that government-funded projects involve higher technical risks is not, of course, necessarily wrong; indeed, this might provide real justification for governmental involvement in the first place, especially for projects of national importance. The main problem governments face lies in identifying high technical risk projects that also have high market potential, and it is doubtful whether, in general, government decision-makers possess the competence to assess market prospects satisfactorily. Thus, we can state that a major problem with innovation policy is the lack of market know-how amongst public policy-makers.

A second area for concern is that of the distribution of government subsidies amongst firms of different sizes. Evidence exists to suggest that by far the greatest proportion of R & D subsidies in most countries has gone to very large firms, precisely those that might be expected to be able to support even major projects themselves. Outside the nuclear, aerospace and electronics industries (which among them have received most of the available government subsidies), funds have generally gone to the support of relatively small, marginal projects of dubious promise that the major companies have been unwilling to finance themselves. Large firms, of course, have considerable political clout as well as the resources to devote to strong political lobbying; small firms have neither of these.

Despite the many resource-related problems that small firms face in innovating (lack of cash resources, lack of R & D manpower, lack of market power), convincing evidence exists to suggest that in many sectors they can make significant contributions to national rates of innovation.[8] Certainly a strong case can be made on social and political grounds, as well as on technological and economic grounds, for governmental support of small firms.[9] A second major problem with innovation policy, therefore, is that of concentration of subsidies in assisting large firms, an imbalance that can and should be redressed.[10]

The fact that governments formulate a battery of measures for the assistance of industrial innovation is, of course, no guarantee that the measures will be taken up on a large scale. There is, in fact, evidence that small firms in particular are largely unaware of many of the government measures that are available to assist them.[11] This suggests that governments should take strong steps to promulgate their measures. Thus, a third problem with innovation policy is that most governments have more often adopted a passive, rather than an active, stance towards information dissemination, policy measures having been taken up largely by a limited number of aware—and usually large—companies.

In the UK, on the other hand, because of active government promotion, the level of awareness of the Department of Industry's microprocessor applications scheme has been relatively high, even amongst smaller firms. The same is true in the related area of robotics. The problem here, however, is that even given awareness of the desirability of adopting robots in production, many small firms simply do not possess the technical expertise to enable them to achieve this satisfactorily. Little is to be gained from leading the horse to water if it is unable to drink! In other words, the provision of awareness without the provision of the necessary skills to enable adoption is simply not enough. Clearly, awareness needs to be coupled to the provision of the requisite training and skills, or at least access to these.

This raises the issue of governmental perception, or lack of perception, of innovation as a multifaceted process. It is a fact that existing governmental policy tools for the stimulation of industrial innovation are mainly on the supply side. In particular, they focus heavily on technical assistance and the financial support of company R & D. In some instances, however, the major costs and often the highest risks have occurred during the post-R & D phases of innovation. More R & D is not necessarily synonymous with more innovation. Further, empirical evidence suggests strongly that innovations fail more often because of poor management, or lack of market awareness or linkages, than as a result of an inability to solve technical problems.[12] Technological competence is an enabling condition for innovation to occur rather than a guarantee of commercial success.

This means that governments committed to the support of industrial innovation should design a co-ordinated package of measures—supply, environmental, and demand—to assist firms' overall innovation endeavours. To achieve this successfully would imply recognition on the part of policy-makers of innovation as an interactive process ranging from R & D, to prototype production, to manufacturing and market start-up, and beyond. A problem with innovation policy in many countries has been a lack of practical knowledge, or imaginative conceptualisation, of the process of industrial innovation by policy-makers, resulting in them adopting a narrow, heavily R & D-orientated, view of innovation to the detriment of other important aspects, such as innovation-orientated public purchasing.

Because different policy tools are normally formulated and often adminstered by separate departments within the government bureaucracy, the need for a complementary set of innovation policy tools imposes the requirements for good interdepartmental co-operation and co-ordination. In practice, of course, these are not always forthcoming. For example, the need to cut public expenditure to reduce budget deficits (treasury departments) does not sit easily with calls for increased investment in strategic technologies or a reduction in corporation tax (industry departments). This can result in internecine bickering in which the national good takes a poor second place. Even given goodwill between departments, lack of co-ordination can result in apparently conflicting initiatives by different departments. This is illustrated in Table 17.6 for the United States. Such a situation is unlikely to occur in Japan, where good co-ordination appears to be the norm. Thus, lack of inter-departmental co-ordination—and sometimes even lack of co-operation—appears to be a major barrier to the formulation and implementation of a coherent innovation policy in many countries. In this respect, the existence of a long-term national strategy towards technological change and economic exploitation might go some way towards facilitating the co-ordination of an otherwise diverse and largely unrelated set of initiatives.

Returning briefly to the question of small firms, there is a notable trend in almost all the advanced market economies towards an intensification of interest in their welfare. This is based largely on a belief in their inherent potential for both innovation and employment generation. By and large, most governments have failed to make any explicit distinction between existing small firms, often in the traditional areas of industry, and new technology-based small firms. It is the latter that have the greatest potential for contributing towards national economic generation. Moreover, small-firm innovation is often a local phenomenon

Table 17.6. Some contradictory policies affecting industry in the United States

The Environmental Protection Agency is pushing hard for stringent air pollution controls.	The Energy Department is pushing companies to switch from imported oil to 'dirtier' coal.
The National Highway Traffic Safety Administration mandates weight-adding safety equipment for cars.	The Transportation Department is insisting on lighter vehicles to conserve gasoline.
The Justice Department offers guidance to companies on complying with the Foreign Corrupt Practices Act.	The Securities and Exchange Commission will not promise immunity from prosecution for practices the Justice Department might permit.
The Energy Department tries to keep down rail rates for hauling coal to encourage plant conversions.	The Transportation Department tries to keep coal-by-rail rates high to bolster the ailing rail industry.
The Environmental Protection Agency restricts use of pesticides.	The Agricultural Department promotes pesticides for agricultural and forestry use.
The Occupational Safety and Health Administration chooses the lowest level of exposure to hazardous substances technically feasible, short of bankrupting an industry.	The Environmental Protection Agency uses more flexible standards for comparing risk levels with costs.

Source: *Business Week*, 30 June 1980, p. 67.

and innovation policy for small firms should thus have a strong local or regional flavour.[13] While there is currently a trend towards regionalisation in the administration of policies towards small firms, policy nevertheless continues to be formulated centrally.

Moving now to more general issues, a desirable property of any innovation policy is that it should be flexible with respect to changing economic and technological threats and opportunities. In other words, policies, once initiated, should not be persevered with unchangingly in the face of changed economic realities. On the other hand, policies should be long term in perspective, reflecting an overall, co-ordinative strategy. Most importantly, they should be divorced from the requirements of party politics, and not be allowed to be altered to suit the often cynical dictates of the two-to-four year political cycle. An important problem with innovation/industrial policy in some countries is that it has often been subjected to major change in accordance with political dogma rather than with changing industrial or economic needs or conditions. One means of introducing flexibility into the operation of particular policy initiatives is to introduce them on an experimental basis. This allows for learning-by-doing, for modification and, if need be, for termination. In this respect, the Experimental Technology Incentives Program of the US Department of Commerce is an interesting example.

An interesting phenomenon in the innovation policy area is the amount of international learning that is currently taking place. Through forums such as the OECD, the EEC and the Six Countries Programme on Government Policies Towards Innovation, awareness amongst governments of what their counterparts in other countries have been, or intend, doing is definitely increasing. This appears to be resulting in some international convergence in the types of policies, or at least policy tools, being adopted. This process does, however, contain inherent dangers. For example, policies that are appropriate in one country, with its characteristic social, political and economic system, might meet with less success in a second country with rather different characteristics and traditions. Care must be taken to tailor the policies and policy tools to suit the adopter country's specific requirements and conditions. In other words, imaginative transfer and interpretation are required.

A second notable area of convergence is that of technology choice. Nearly all recent governmental policy statements contain reference to building up strengths in a handful of technology areas, notably microelectronics, biotechnology, and information technology. Clearly, all countries cannot be across-the-board winners with all three technologies. It is remarkable how many countries—even relatively small ones—are determined to construct large-scale microelectronics components industries, apparently completely ignoring the determination of their neighbours to do likewise. Can the smaller western economies hope to compete in full-scale microelectronics production with the two industrial giants, Japan and the US? For most—with the possible exception of the Netherlands—the answer must be 'no'. The solution for the smaller countries must surely be to go for the development of specialist devices, or to concentrate on the use of microelectronics in the production of new products and processes. In fact, in all three of the above technologies, for the smaller economies, a niche strategy would appear to offer the greatest—and perhaps the only—prospect of success.

Finally, having discussed innovation policy in some detail, it must be admitted that we know very little about its effectiveness. It is a fact that few innovation policy initiatives have been subjected to objective assessment of their efficacy. There is thus a pressing need for detailed impact assessment across a wide range of policies and in many countries.[14] In view of this, it might be conjectured that perhaps the greatest problem of innovation policy is that it has been more an object of faith rather than of understanding.

Despite this, innovation policies are an inescapable fact of modern economic life. With increasing interdependency in world trade, no country can afford to isolate itself from developments elsewhere and opt out from the need to stimulate, assist with, and even guide, national innovative activity. If, as a number of eminent economists believe, technological change is the key to the next economic upswing, then the need for effective, imaginative and far-reaching national innovation policies is even more pressing.[15] It seems likely that, despite past problems, innovation policies can be made to work.

Notes and references

1. Nelson, R. and S. Winter (1977), 'In search of a useful theory of innovation', *Research Policy*, 6 : 36-76.
2. Rothwell, R. and W. Zegveld (1982), *Innovation in the Small and Medium Sized Firm: Their Role in Employment and in Economic Change*, London, Frances Pinter.
3. Rothwell, R. (1980), 'The impact of regulation on innovation: some US data', *Technological Forecasting and Social Change*, 17 : 7-34; Rothwell, R. (1981), 'Some indirect impacts of regulation on innovation in the United States', *Technological Forecasting and Social Change*, 18 : 57-80.
4. Jewkes, J. (1972), *Government and High Technology*, Occasional Paper No. 37, London, Institute of Economic Affairs; Clark, R. (1973), 'Mintech in retrospect', *Omega*, 1 (1): 25-38; 1 (2): 137-63.
5. Sciberras, E. (1980), 'The UK semiconductor industry' in Pavitt, K. (ed.), *Technical Innovation and British Economic Performance*, London, Macmillan.
6. Gardner, N. (1976), 'Economics of launching aid' in Whiting, A. (ed.), *The Economics of Industrial Subsidies*, London, HMSO; Little, B. (1974), *The Role of Government in Assisting New Product Development*, School of Business Administration, University of Ontario, Working Paper No. 114, London, Canada.
7. Walker, W. (1976), *Direct Government Aid for Industrial Innovation in the UK*, report to Organisation for Applied Scientific Research, Delft, Netherlands, mimeo, Science Policy Research Unit, University of Sussex.
8. Rothwell and Zegveld, op. cit.; Townsend, J., F. Henwood, G. Thomas, K. Pavitt and S. Wyatt (1981), *Science and Technology Indicators for the UK: Innovations in Britain since 1945*, Science Policy Research Unit, Occasional Paper No. 16, University of Sussex; De Melto, D., K. McMullen and R. Wills (1980), *Innovation and Technological Change in Five Canadian Industries*, Economic Council of Canada, Discussion Paper 176, Ottawa.

9. Rothwell and Zegveld, op. cit.
10. Notable exceptions are the Netherlands and Denmark, where small firms appear to receive a fairer slice of the government-subsidy cake.
11. Rubenstein, A. *et al.*, (1977), 'Management perceptions of government incentives to technological innovation in England, France, West Germany and Japan', *Research Policy*, **6** : 324.
12. Rothwell, R. (1977), 'Characteristics of successful innovators and technically progressive firms', *R & D Management*, **7** (3): 191-206.
13. Rothwell, R. (1982), 'The role of technology in industrial change: implications for regional policy', *Regional Studies*, **16** (5): 361-9.
14. Goldberg, W. (1981), *Exploration into the Instrumentality of Innovation Policy*, International Institute of Management, Science Centre, Berlin.
15. Freeman, C., J. Clark and L. Soete (1982), *Unemployment and Technical Innovation*, London, Frances Pinter.

Author Index

Abramovitz, M., 17, 24n
Aldcroft, D., 35n
Alderson, W., 34n
Amin, S., 153n
Anderman, S., 190n
Andrews, P., 80, 91n
Arendt, H., 191n
Arnon, N., 74n
Arrow, K., 1-2, 3n, 75, 58, 83, 84, 88n, 91n, 92n, 177n

Babbage, C., 13, 24n
Backhouse, R., 89n
Badger, G., 36n
Bain, A., 55n
Barron, I., 176n
Baxter, P., 36n
Beales, H., 89n
Begg, D., 78, 91n
Bell, D., 75, 88n
Bennett, E., 176n
Berg, I., 190n
Bhagwati, J., 83, 91n
Bhalla, A., 35n
Bishop, E., 35n
Bjørn-Andersen, N., 90n
Blackaby, F., 25n, 133n
Blumenthal, T., 167n
Bonnen, J., 89n
Botkin, J., 92n
Boulding, K., 75, 88n
Bradburd, R., 79-80, 89n, 91n
Brandt, W., 154n
Braun, E., 35n, 176n
Bridges, T., 24n
Brooks, H., 191n
Brown, A., 176n, 177n
Brown, L., 47, 49n, 141-2, 154n
Brown, R., 34n

Cannan, E., 90n
Carstairs, R., 167n, 168n
Carter, C., 133n
Castano, A., 61, 72n, 73n
Caves, R., 89n

Chamala, S., 89n
Chandler, A. D., 92n
Chapman, D., 28, 34n
Chee-Wah Cheah, 89n
Chow, G., 95, 103n
Clark, J., 215n
Clark, R., 214n
Coase, R., 90n
Cockcroft, D., 190n
Coffey, M., 176n, 191n
Cole, G., 189n
Colman, D., 153n
Compaigne, B., 92n
Contractor, F., 159, 167n, 168n
Cornwall, J., 133n
Correa, C., 168n
Craswell, R., 89n
Crawford, N., 163, 168n
Crompton, R., 189n
Crookwell, H., 89n
Crouch, B., 89n
Cunningham, W., 11, 12, 24n
Curnow, R., 176n

Dahlman, C., 58, 72n, 73n
David, P., 93, 97, 98-9, 101, 103n
Davie, M., 3n
Davies, H., 164-5, 167n, 168n
Davies, S., 93, 96-7, 99-101, 102, 103n
De Melto, D., 214n
Denison, E., 17, 24n, 174, 177n
Dewey, D., 49n
Diaz-Alejandro, C., 153n
Dunn, D., 76, 90n
Dunning, J., 154n
Dunphy, D., 176n, 191n

Eade, D., 89n
Earl, M., 90n
Eckholm, E., 152, 154n
Eilon, S., 121n
Eliasson, G., 176n
Elliott, E., 11, 23n
Elliott, R., 36n
Elston, J., 92n

218 *Author index*

Enos, J., 57, 67, 72n
Evans, O., 177n

Fabricant, S., 17, 24n
Fischer, A., 103n
Fonseca, F., 58, 72n, 73n
Ford, W., 176n, 191n
Forward, P., 176n
Freeman, C., 175n, 177n, 215n
Freeman, Marcia, 190n
Freeman, Michael, 190n

Galatin, M., 89n
Galenson, W., 144, 154n
Gallopin, G., 55n
Gannicott, K., 31, 35n
Gardner, N., 214n
Gertler, M., 89n
Gibbons, M., 35n
Glaister, S., 94–5, 103n
Gold, B., 121n
Goldberg, R., 89n
Goldberg, W., 215n
Gomulka, S., 128, 132n, 133n
Griliches, Z., 72n
Grossman, S., 89n

Hahn, F., 91n
Håkansson, H., 168n
Hayek, F. von, 75, 78, 88n
Heertje, A., 49n, 130, 133n
Helleiner, G., 153n, 158, 167n, 168n
Henwood, F., 214n
Hirchleifer, J., 77, 78, 87, 89n, 90n, 92n
Hodge, B., 176n, 177n
Hodgskin, T., 76, 90n
Hodgson, J., 89n
Hollander, S., 57, 67, 72n
Hollingdale, S., 35n
Holst, O., 90n
Hudson, K., 24n

Ireland, N., 101, 102, 103n
Ironmonger, D., 55n

Jenkins, C., 190n
Jensen, R., 34n
Jewkes, K., 214n
Johnston, R., 35n
Jonscher, C., 83, 91n
Jorberg, L., 24n
Jussawalla, M., 89n, 153n, 154n

Kamien, M., 47-8, 49n
Katz, J., 72n, 73n
Kennedy, C., 133n
Killing, J., 89n, 168n

Kirkland, R., 153n
Kitamura, H., 153n
Klapp, O., 36n
Klein, B., 138, 153n
Klein, L., 132n
Knight, F., 75, 77, 78, 86, 88n, 90n
Kornai, J., 75, 88n
Kornhauser, C., 177n
Kramer, R., 131, 133n

Lamberton, D., 34n, 35n, 36n, 88n, 89n,
  90n, 91n, 92n, 153, 153n, 154n, 176n,
  177n, 191n
Leibenstein, H., 144, 154n, 170, 172-3,
  176n, 177n
Leiter, R., 89n
Leonard-Barton, D., 35n
Levine, A., 189n
Lindner, R., 97, 103n
Little, B., 214n
Livingston, J., 2, 3n
Livingstone, I., 153n
Lorin, H., 89n
Lowe, J., 163, 168n
Lundberg, E., 133n

McArthur, E., 12, 24n
MacBride, S., 154n
McCain, R., 91n
McCall, J., 90n
McColl, G., 176n
Macdonald, S., 34n, 35n, 36n, 91n, 92n,
  154n, 176n, 177n
Machlup, F., 75, 77, 88n
Mackaay, E., 90n
MacLaughlin, R., 177n
McMullen, K., 214n
Magdoff, M., 137, 153n
Magee, S., 137, 153n
Mandeville, T., 34n, 35n, 36n, 91n, 92n,
  154n, 176n, 177n
Mansfield, E., 35n, 55n, 89n, 93-5, 97, 103n
Mansfield, W., 190n, 191n
Manwaring, A., 191n
Marris, R., 128, 132n
Marschak, J., 75, 88n, 177n
Marshall, A., 24n
Martin, J., 24n
Martino, J., 121n
Mathias, P., 24n
Mauro, K., 89n
Maxwell, P., 72n, 73n
Mensch, G., 133n
Metcalfe, J., 95-6, 103n
Mumford, E., 90n

Nabseth, L., 109, 121n, 133n

Navajas, F., 72n, 73n
Nelson, R., 3n, 34n, 36n, 87, 90n, 91n, 92n, 153n, 202, 214n
Nixson, F., 153n

O'Driscoll, G., 91n
Oettinger, A., 92n
Ohkawa, K., 132n
Oravainen, N., 161, 167n, 168n
Over, A., 79–80, 89n, 91n

Pardey, P., 103n
Parker, W. N., 24n
Pavitt, K., 214n
Peirce, W., 121n
Pengilley, W., 168n
Perlman, M., 89n
Pommerehne, W., 89n
Porat, M., 90n, 177n
Prebish, R., 136, 153n
Price, D., 92n
Price, D. de Solla, 36n

Ramos, A., 92n
Ranis, G., 154n
Ray, G., 55n, 109, 121n, 133n
Reinecke, I., 36n
Reinganum, J., 102, 103n
Riemenschneider, C., 89n
Riley, J., 77, 78, 87, 89n, 90n, 92n
Robbins, Lord, 76, 90n
Robinson, A., 35n, 153n
Rogers, E., 35n
Romeo, A., 102, 103n
Ronstadt, R., 131, 133n
Rosegger, G., 121n, 176n
Rosenberg, N., 16, 17, 18, 24n, 25n, 55n, 73n
Rothwell, R., 70, 72n, 74n, 133n, 176n, 203, 205, 206, 207, 208, 209, 214n, 215n
Rubenstein, A., 215n

Salop, S., 89n
Salter, W., 17, 24n
Sargent, J., 25n
Sassone, P., 92n
Sawyer, G., 35n
Scherer, F., 35n
Schiller, H., 137, 153n
Schmookler, J., 24n, 72n
Schneider, F., 89n
Schotter, A., 80–1, 91n
Schuman, P., 90n
Schumpeter, J., 1, 3n, 34n, 35n, 123, 132n, 176n
Schwartz, N., 47–8, 49n
Sciberras, E., 214n
Segafi-Nejad, T., 167n
Senker, P., 176n
Shackle, G., 75, 88n
Sherman, B., 190n
Simon, H., 75, 79, 84, 88n, 91n, 92n
Slater, J., 91n
Smiles, S., 13, 14, 24n
Soete, L., 174, 177n, 215n
Solow, R., 17, 24n
Sowell, T., 90n
Spence, A. M., 77, 88n, 90n, 95, 96, 102, 103n
Stanley, M., 1, 3n
Stanley, P., 172, 177n
Starrett, D., 91n
Stewart, F., 145, 154n, 156–7, 167n
Stiglitz, J., 77, 89n, 90n
Stoneman, P., 95, 97, 101, 102, 103n
Stopford, J., 167n
Stout, D., 91n, 133n, 176n
Strassman, P., 175, 177n
Streeten, P., 145, 154n
Svennilson, I., 133n

Teece, D., 34n, 161, 167n, 168n
Teubal, M., 62, 72n, 73n, 74n, 80
Thomas, G., 214n
Thompson, D., 168n
Thornton, B., 172, 177n
Tilley, R., 121n
Tiltman, J., 35n
Tinbergen, J., 91n
Tomasini, L., 88n
Toothill, G., 35n
Townsend, J., 214n
Toynbee, A., 11–12, 24n
Trachtenberg, M., 74n
Tunzelmann, G. von, 14, 24n
Turnbull, P., 163, 168n

Uhlmann, L., 110, 121n
Ure, A., 13, 24n

Walker, W., 210, 214n
Ward, B., 152–3, 154n
Warner, G. Townsend, 12–13, 24n
Wedderburn, D., 189n
Weizenbaum, J., 177n
Welch, L., 167n, 168n
Welch, R., 90n
Wells, L., 167n
Wiedersheim-Paul, F., 168n
White, L., 154n

Williams, B., 77, 90n
Wills, R., 214n
Winter, S., 3n, 36n, 87, 90n, 92n, 202, 214n
Withington, F., 177n
Wolfe, S., 91n

Wootz, B., 168n
Wyatt, S., 214n

Zegveld, W., 133n, 176n, 203, 205, 206, 207, 208, 209, 214n, 215n
Zimmerman, R., 139, 141, 153n, 154n

Subject Index

ACARD, 205
Advertising, 95
APEX, 179, 184–5, 189–90
Argentina, 21, 143, 148
Art and technology, 2
Artificial intelligence, 78
ASTMS, 179
Australia, 6, 30, 132, 180, 194
Australian Atomic Energy Commission, 32
Austria, 180

Belgium, 180
BIFU, 179, 191
Biotechnology, 124
Brandt Commission, 150
Brazil, 21, 127, 132, 139, 143, 148
Britain, 18, 21, 22, 129
British Rail, 16

Canada, 135, 139, 203, 205, 208, 209
Capital
    accumulation, 2, 56–71, 80–81
    human, 17
    organisational, 2, 79–80
Capital/labour ratios, 144
Caracas Conference, 151
Catholic Church, 4
Chemical bleaching, 12
Chile, 143
Cobb-Douglas production function, 44
Cognitive simulation, 79
Colombia, 142
Communication, 161–2, 193–201
Comparative advantage, theory of, 124–7
Competition
    and innovation, 104, 130
    dynamic, 129
    foreign, 19
    international, 122–132
    Schumpeterian, 122–3
Computational complexity, 78
Computers, 26, 30, 196–201
Concorde, 18, 125
Constant elasticity of substitution (CES), 44, 95

Costs, 33–4, 66–7, 79, 86–7, 98, 105–20, 174–5
'Countertrade', 138–9
Crystal Palace, 7

Decision criteria, 110–20
De-industrialisation, 189
Demand, 15, 130
Denmark, 19
Developing countries, 134–53
Development economics, 16
Diffusion of technological change, 15, 37–8, 41, 47, 43–4, 93–103, 105–10, 125, 130, 149
DNA, 4–5, 130

Ecological system, 5
Economic growth, 14, 123–4, 128
Economic historians, 11, 15–16
Economic policy, 55, 202, 205
Economic theory, 3, 16, 37–8, 44–6, 48, 50–1, 83, 124, 136
Economies of scale, 19, 122, 131
Education, 17
EEC, 189, 213
Employment
    and technological change, 1, 27, 32, 171–5
    industrial growth and, 12
    in information occupations, 81–2
Endogenous technological change, 48, 123, 130
Entrepreneurship, 98
Entry, conditions of, 123–4
Equity, 142
European Trade Union Institute, 180
Evolution, 4–5
Exogenous technological change, 41, 48, 130
Export-Import Bank, 135
Exports, 122–32

Factor price equalisation theorem, 123
Factor substitution, 144
FIET, 180, 191

Subject Index

Firms, 91, 102–3, 108
Flying shuttle, 11
Forecasting, 27, 188
France, 21, 22
Frascati Manual, 30
Free trade, 126, 140

General Agreement on Tariffs and Trade (GATT), 138–9
General equilibrium theory, 83
General Motors, 4

Heckscher–Ohlin theorem, 122
Hong Kong, 126, 149

ICFTU, 180
Industrial archaeology, 16
Industrial recolution, 7, 16, 170
ILO, 180
Imports, 150
India, 21, 132, 148, 164
Indian Space Research Organisation, 145
Informatics, 139, 142
Information
  and innovation, 33–4
  and uncertainty, 75, 77
  costs, 33–4, 79, 86–7, 98
  distribution, 82, 87
  economics, 75–88
  economy, 81–2
  elasticity of demand for, 85
  flows, 30–1, 193–201
  infrastructure, 92
  investment in, 76
  processors, 82
  producers, 82
  public good aspects of, 86
  sector, 81–3
  stock, 80
  technology as, 2, 27, 87–8
  underutilisation of, 85–7
Informational efficiency, 76
Innovation
  and information, 33–4, 126
  and market structure, 47–8
  and patent activity, 129–30
  concept of, 15, 105–20
  diffusion of, 37–8, 41, 47, 53–4, 93–103, 105–10, 125, 130
  linear model of, 28, 31
  policy, 202–13
  process, 29–31, 34, 52, 70–1, 105–20, 132
  subsidies, 209–10
Intangibles accumulation, 2, 56–71, 80–1
International Development Agency, 135
International Metalworkers' Federations, 180

International Monetary Fund, 135
International trade, 122–32, 135–53
Inventions, 12–13
Italy, 180

Japan, 130, 131, 138, 144–6, 149, 156, 180, 204–5, 208–9, 213
Japanese Fifth Generation Computer Project 76

Kenya, 144–5
Knowledge, 4, 10, 76–7
Korean Institute of Science and Technology, 145

Labour
  displacement of, 1, 12, 172–5, 179, 182–8
  productivity, 19, 27
  supply, 146
Labour saving techniques, 171–5, 190–1
Language, technological, 193–201
Leading sectors, 15
Learning process, 59, 93–7, 102
Licensing, 131, 155–67
'Light-footedness', 129
Linear innovation model, 28, 30–1

MacBride Commission, 152
Madras Leather Institute, 145
Machine-making, 12
Management information systems, 33
Management science, 78
Manhattan Project, 1
Manufactures, 11
Market economy, 1, 76
Market structure, 47–8
Mechanisation of transport, 12
Metalworking, 60–2
Mexico, 142, 145, 148
Microelectronics, 125, 169, 179, 191
MITI, 92
Monopoly, 147, 169
Mule, 11–12
Multinational firms, 1, 26, 131, 146, 161

Neolithic, 7
Netherlands, 204–5, 208–9, 213
Newly industrialising countries (NICs), 127, 148–9
New Zealand, 180
NIEO, 134
Non-tariff barriers, 138
Nordic Council of Trade Unions, 180
Norway, 185

Oligopoly, 103, 122, 138

OPEC, 139
Operations research, 79
Organisation
  as a variable, 84
  as capital, 2, 79–80
  economics of internal, 3, 84–5, 163–5
  role of, 18, 20, 87, 160
Organisational
  artefacts, 4
  change, 79–80, 170–1
  obsolescence, 3, 78

Palaeolithic, 4, 6
Patents, 129–30, 158–61
People's Republic of China, 148
Perfect knowledge, 2, 78
POEU, 179
Policy
  formation, 104
  innovation, 202–13
  issues, 166
  science and technology, 32–3
  trade-impeding, 135
Pottery, 12
Power loom, 11–12
Probit models, 97–102
Process industries, 57–60
Process innovations, 51–2, 93
Product innovations, 51–2
Project SAPPHO, 56, 74
Production functions, 39–46, 72, 123–4, 147
Productivity, 15–18, 27, 136, 170, 174–5, 185–7
Profitability, 62–9, 73, 96, 108, 130
Property rights, 15, 78
PTTI, 180

Rationality, 79
Reciprocity, 138
Research and Development (R&D)
  and information activities, 77–8, 83
  and industry structure, 20, 145
  and intangibles accumulation, 66–7, 71
  and patenting, 129–30
  and technology, 20, 56, 87, 130, 211
  costs of, 66–7, 105–20
  government support of, 30, 202–14
  statistics, 30
  underinvestment in, 145

Savings, 110–20
Scale, 19, 122, 131
Science and technology policy, 32–3, 202–14
Service sector, 19–23
Shipping, 21
Singapore, 149

Six Countries Programme on Government Policies, 213
Skills, 129, 139, 141, 163
South Korea, 126, 127, 142, 144, 145, 148, 149
Spinning jenny, 11
Steam engine, 11–12
Steam power, 13–15
Structure change, 23, 188–9
Substitution, elasticity of, 44
Sweden, 19, 21, 181, 204, 205, 208, 209
Switzerland, 19, 180

Taiwan, 127, 142
Technological change
  and cost reduction, 57–9
  and economic growth, 14, 18
  and economic theory, 16, 37–8, 44–6, 50–1
  and employment, 1, 12, 32, 172–5,, 179, 182–8
  and productivity, 15–18, 27, 136, 170, 174–5, 185–7
  and trade, 122–32
  as information flow, 30–1
  diffusion of, 37–8, 41, 47, 53, 93–103, 105–10, 125, 130, 149
  empirical studies of, 105–10
  endogenous, 48, 123, 130
  exogenous, 41, 48, 130
  information, 169–75
  modelling process of, 50–1, 71
  neutral, 124
  predictability of, 37–8, 46
  process of, 33–4, 43–4
  trade union view of, 180–2
Technology
  as information, 30, 87–8, 156–8
  as machines, 26
  changing, 50–1
  defined, 7, 26–7, 50, 52, 169, 190–1
  fixed, 50–1
  for economic management, 55
  office, 185
  transfer, 129–32, 135–53, 155–67, 179
  trouble with, 1, 193
  without machines, 153
Telecommunications, 29, 132, 169
Trade, 1, 122–32, 135–53
Trade Union Advisory Committee of OECD, 180
Trade Unions, 178–89
Transfer pricing, 137
Transnational corporations, 137
TUC, 179, 191

Uncertainty, 103, 108

Subject Index

Unemployment, 1, 12, 172-5, 179, 182-8
UNIDO, 146
UNISIST, 85
United Kingdom, 22, 179, 204, 205, 208, 209, 211
United Nations, 4, 140
United States, 21, 127, 131, 135, 138, 149, 204, 205, 208, 209, 212, 213
Uruguay, 143

Vintage model, 98

VLSI circuitry, 131

Water frame, 11
West Germany, 21, 22, 180
Wheat Research Institute, 145
World Bank, 135
World Intellectual Proprty Organisation, 138

X-efficiency, 172-3
'Xerox' effect, 83, 174-5